TRAINERS MANUAL

TRAINING TECHNIQUES OF
Creative
Problem Solving

R.G. CHAUDHARI

INDIA • SINGAPORE • MALAYSIA

Notion Press

Old No. 38, New No. 6
McNichols Road, Chetpet
Chennai - 600 031

First Published by Notion Press 2018
Copyright © R.G. Chaudhari 2018
All Rights Reserved.

ISBN 978-1-64324-132-6

1

TRAINING TECHNIQUES OF CREATIVE PROBLEM SOLVING

CONTENTS

PREFACE

This manual is designed to be a spring board for a voyage to the frontier of creative thinking. An attempt is made to demonstrate that Creative Problem Solving (CPS) is a proven method of solving problems in an innovative way. Efforts are made to convince the participants of CPS workshop and other readers to realize that creativity can be applied each day to many aspects of our every-day life. Inputs of the workshop will constantly remind them that the goal of assimilating the knowledge of CPS is to continue to develop their own creativity, in their own way and to move forward by fallowing the leads.

Being an initiative programme, no attempt is made to make it a comprehensive approach to understanding how human systems work. It is simple and full of fun.

Methodology of presenting the contents of the course is based on time-tested training techniques. They have proved useful to me in my job over two and half decades in a heavy engineering industry, as a trainer, on the shop floor and equal period in running my own small scale industry.

As regards the contents of the course, all important dimensions of creativity are illustrated by several exercises collected from the published literatures, and many of them are based on my own experience. In fact, the experience that I gained stems from participation in courses conducted by Certified Value Specialists (CVS), Institute of Creative Development, Society of Indian Value Engineering, CraVe (Bangalore), etc. Knowledge acquired by being a Member of Creative Education Foundation, USA, has also contributed in my efforts to put the techniques of CPS in a simple way.

ACKNOWLEDGEMENTS

I have freely drawn from the inputs provided by Indo-American Society, Bombay (Mumbai), Institute for Creativity Development, Poona (Pune) and National Productivity Council, New Delhi, during the programmes conducted by them. My thanks are due to them for inspiring me to compile this manual.

I acknowledge my gratitude to the Management of Bharat Heavy Electricals Ltd. For giving me many opportunities to participate in Value Engineering and CPS Workshops, and to conduct numerable Executive and Supervisory Development programmes in BHEL. I owe my thanks to Creative Education Foundation, USA, for enrolling me as a Member and for supplying their publications like 'Occasional Papers.'

I am indebted to my erstwhile colleagues in Bharat Heavy Electricals, Hyderabad's Human Resource Development Centre, Productivity Services, Prototype Development Centre and also Trainees/ Participants in the programmes conducted by me, who have inspired me to take up the work on this manual.

Hyderabad-India   R.G. Chaudhari

January 2018

OVERVIEW OF CPS
AN EXPLORATION IN CREATIVE THINKING PROCESS

This short course provides an invigorating experience to the participants and improves their problem solving skill through better utilization of latent creative potentials… a pleasant encounter with the creative thinking process which keeps them wondering why they did not learn it 10 or 15 years ago.

Why Creative Problem Solving?

The astonishing strides made by man from pre-historic days to space-age bear eloquent testimony to the power of human thoughts. Mind-boggling Technological advances achieved by many nations owe their incredible success to the creative talent of their human resources. The leaders of successful enterprises have maintained undoubted faith of the power of human mind. Every effort is, therefore, made to develop that potential as an on-going exercise in generating novel and worth-while ideas for solving all sorts of problems by newer, innovative techniques.

Back at home, we look up the hierarchical ladder for judicial judgment and conventional wisdom in solving problems. In doing so, little is done to cultivate and nurture creative potentials of human assets. We need now, more than ever before, education in innovating thinking to enable us to achieve our potential development as individuals and as a nation.

This workshop aims just at that objective. It is designed to act as a spring board in awakening the dormant talent, leading to creative action in professional and personal life as a force of habit.

What CPS aims to Achieve?

Strolling along the shores of creativity, CPS seeks to reveal the bitter truth that our creative thinking ability is dormant, waiting to be aroused from its slumber and that we have a much greater thinking capacity than ever we use.

While unfolding this truth, discussions bring out strategies of removing surface barriers and inhibitions that stunt natural growth of imaginative faculty, through practical examples and personal experiences. After getting the feeling of release, discussion leads to ways and means of developing creativity.

What Questions CPS Seeks to Answer?

There are three key words in CPS: Namely: (1) Problem (2) Solution and (3) Creativity

1. **Problem:** What is a problem?

   Can you define it precisely?

   Where do you find problem?

   Do you really solve a problem or create more problems in the process of solving one problem?

2. **Solution:** What is a solution?

    How many methods of finding solution?

    Which method is better and why?

3. **Creativity:** What is meant by creativity?

    What are the principal abilities of human mind?

    Are intelligence and creativity synonymous?

    Are you creative?

    No doubt about it but do you know that not more than 10% to 12% of it is put to use?

    Why?

    Is it because of race?–sex?–age?–education?–environment?

    Is creativity in-born or inspirational?

    Can our creative output be increased?

    Are there any deliberate methods?

    What are they?

    What are the traits of a creative person?

    What factors foster creativity and what stifles free flow of ideas?

    In short, how to improve our creative output?

Who Should Attend?

Decision makers & organization 'shapers,' as well as individuals, who would like to get out of their mental rut and unlock their creative potentials. In other words:

–   Who are dissatisfied with routine methods of handling problems.

–   Who are ready to look at problems as opportunities.

–   Who have will and desire to excel in resourcefulness.

–   Who need to recognize and tap the creative power of their companions/associates.

–   In brief, who desire to improve quality of life – on and off the job.

2

TRAINING TECHNIQUES OF CREATIVE PROBLEM SOLVING

GENERAL GUIDELINES:

1. FACILITIES & ARRANGEMENTS

2. HINTS FOR THE TRAINER

3. MECHANICS OF PRESENTATION

4. USING THE MANUAL

5. SESSION-WISE THING YOU NEED

6. SUGGESTED TIME SCHEDULE

1. FACILITIES & ARRANGEMENTS

1.01   Well ventilated hall large enough to accommodate about 16 to 20 members seated in conference style and a row of about 12 extra chairs along the wall…

1.02   A chalk board of about 4' × 5' size or a rolling type chalk board, ample supplies of white and colour chalks and 2 numbers of good quality dusters, are indispensible.

1.03   There are a number of good films on creativity are available. Efforts should be made to screen at least two such films during the workshop. Film projector either provided by sponsor or Trainer's own should be in good working condition.

1.04   The suggested flip charts and transparencies for over head projection should be prepared in advance, with a touch of an artist. (Once made, they will be useful for many programmes).

1.05   An easel for charts, a slide projector, an overhead projector and any other electronic devices that are planned to be used, should be tested at least one day in advance and kept ready for operation at the flick of a switch.

1.06   Handouts should be kept in a side board, in sequence of their distribution.

1.07   Name strips of the size 20 × 80 mm with names typed on computer and laminated that can be clipped to shirt packets, should be kept ready one day in advance for distributing to participants and guests.

1.08   Efforts should be made to arrange an exhibition of Creative Problem Solving (CPS) projects using charts, photos, models and real jobs. Such exhibits make most effective and positive impact on the minds of observers.

1.09   Arrangement for serving tea/coffee on time should be checked.

NB: The following clauses and also clause No 2.11* are applicable if the propramme is sponsored by an organization.

1.10   Sufficient number of copies of the speeches to be delivered by the representative(s) of management and the guest(s), must be made available to the audience before the start of the session(s).

1.11   It should be ensured at least one day in advance that the nominated person(s) is (are) attending the programme and will be in his (their) seat(s) 15 minutes before the start of the inaugural session.

1.12   It is equally important that the invitations to the invitees are in their hands 2 or 3 days before event.

1.13   One of the time tested technique of motivating the participants to give their best is to promote healthy competition, and award prizes or rolling shield to the best team. If this idea finds approval of the management, the prizes/shields should be exhibited in the hall on the inaugural day with some catchy slogans kept underneath them.

***

2. HINTS FOR THE TRAINER

2.01   Be in the hall 15 minutes before the start of the session.

2.02   Check the facilities and handouts required for the session are in their places, in good condition.

2.03   Put the participants in relaxed mood with a friendly word as they arrive.

2.04   Never comment on personal attire or coming late of any member.

2.05   Be cheerful and enthusiastic. Your attitude and mood is the trend setter for the others. Please remember programme can be no way better than you.

2.06   Know your manual thoroughly, which can be done by reading it several times.

2.07   Vary presentation to suit the group: Change the wordings and examples to suit the composition of the group, and use the manual as a guide.

2.08   <u>**Watch your timing: Start and finish promptly. Do not treat each session as a complete entity. If time permits, bring forward the first portion of the next session. If you are short of time, defer the leftover portion of the session to the next seating. But do not omit any portion at any time.**</u>

2.09   Do not attempt to 'lecture' or domineer. Any attempt to make the participants to accept your view will fail. Respect their views and feeling, and steer clear controversies.

2.10   Follow the well-known teaching etiquettes, a few important amongst them being the followings:

   a.   Plan your chalk board work: Make full use of the space as suggested in the 'Procedure' column of the manual.

   b.   Do not over write, obliquely or haphazardly.

   c.   Write in your natural style but ensure that writing is legible.

   d.   Do not present your back to the audience while writing on the board.

   e.   Talk while writing on the chalk board but raise voice.

   f.   Do not look at one or two members at all times but look around the whole group, and speak in well-modulated voice.

   g.   Avoid actions like rolling a shirt button in fingers, chewing something in mouth or (removing your spectacles every now and then), shaking legs, drumming of fingers, etc. Such actions distract the attention of the participants and irritating to the viewers.

   h.   Also avoid repeated use of expressions like 'you see,' 'you know,' 'catch my point.'

   i.   To avoid monotony and boredom, vary your tone frequently, and do not talk continuously for more than 15 minutes. The manual provides necessary breaks at appropriate intervals.

2.11*   Extend help in preparation of inaugural and valedictory addresses by the head of the organization/chief guest. This can be done in two ways:

a. Preparing the script and submitting for their/his perusal and modification.

b. Listing out all the salient features of CPS and their relevance to today's situation in organization.

The inaugural speech by the head of the organization should contains, besides importance of the workshop, enough hints to the participants that the management is seriously interested in the workshop, that the participants should ensure their attendance in all the sessions and that the techniques should be practiced as a part of their normal duties.

***

3. MECHANICS OF PRESENTATION

3.01   Discussion Leading

The manual relies heavily on group discussion. Therefore, success of this technique lies in getting full participation of the members and maintaining correct line of discussion. This calls for thorough preparation on the part of discussion leader i.e. You. It is, therefore, essential that you should know the manual thoroughly, give previous thought to the subject matter and understand the main and subsidiary questions to be posed to reach the desired conclusion. It is equally important to have a large fund of personal examples, particularly from the industries or professions to which the participants belong, to illustrate the points under discussion.

Audio-visual aids, even the chalk board itself and demonstrations from you and the participants, are the aids for discussion, and you should be able to choose and use best of them at appropriate time.

3.02   Opening the Discussion

You should open the discussion by giving a brief introduction to the subject, stating a few facts and quoting well known opinions. These statements should lead discussion on proper lines which will be evident from the responses to your questions.

3.03   Leading the Discussions

Once the discussion gets going, it is your job to stimulate and guide each member in clear thinking and understanding. You should not discourage any talkative member or directly attack a less articulate member as if he is not showing any interest in the deliberations. Discussion from the talkative participant can be moved from him by asking a direct question to some other member. On the other hand, the less articulate member can be encouraged to participate in the discussion by directing question(s) to him. In fact, there are many ways of doing this and you should use own judgment to suit the situation.

You should keep close watch on faulty reasoning or sweeping generalization such as

'Nobody is interested in development work in my company' or 'you cannot get work done from the workers these days.'

You should avoid use of jargon and be prepared to clarify the meaning by re-wording the statement in simple language. Be in readiness to distinguish between fact and opinion;

objective and subjective statements. It may be a fact that I started the session 10 minutes after the schedule time, but to state that I stated the class 'very late' is a matter of opinion. A fact is not open to argument but its interpretation can be a subject for discussion. Many a time you will be tempted to give your own opinions or views but that is very risky. It may involve you in endless argument or futile cross-talks.

3.04 **Controlling the Trend of Discussion**

If the discussion trends to narrow down to a single point of view, you should ask leading questions and pave the way for opening wide the field of discussion to accommodate other points of view. However, you should ensure that discussion is always kept and run on the track but not dragged. If it appears to wander, firmly but tactfully bring it back to the subject. This can be done by telling the member that 'you appear to have a good point there but presently we are discussing "………………." or "Mr. …… seems to make a useful contribution. May I request him to remind me about that point at appropriate time." Such statements usually have a desired effect.

Notice how the participants change their opinions as the discussion proceeds. Without making direct reference to such changes, use them to help you in reaching the desired conclusions more quickly.

3.05 **Summing up**

Once a point is discussed in all aspects, you should review the point and sum up the conclusions reached. The summary should be objective and leave the group with clear idea of what has been achieved.

3.06 **Handling Participants' Questions**

a. Deal with one question at a time.

b. Do not allow interpretation or rewording of a member's question by other members.

c. Discourage interruptions and arguments, tactfully.

d. Understand the question fully by getting the facts cleared.

e. Pass the question to the group, making sure that they understand as intended.

f. Summarize the group opinion as a reply to member's question.

***

4. USING THE MANUAL

4.01   To use the manual successfully, you must understand the portion on which you are working so that the main theme can be <u>stressed</u> in several ways: e.g. by inflexion of voice, or use of deliberate pauses or exaggerated facial expressions and body action.

This requires good practice which comes by going through the manual several times and practicing the methods of <u>stressing</u> in privacy. While using the manual, use pauses to glance down and get the sense of next sentence or paragraph in the 'statement' column. Do not readout word by word from the manual. In fact, let your manual be inconspicuous.

4.02.   The 'Procedure' column in the manual gives details of or clue to the thing to be done by you; and the 'Statement' column tells you what you should state or describe. Both these parts follow a definite sequence. Any slip or mix-up will lead to confusion, something like putting the shoes first and then wondering what to do with the pair of socks. To avoid such embarrassment, I repeat: read your manual thoroughly and organize your thoughts sequentially.

4.03   The manual is fully loaded with exercises and examples.

The best way of presenting the essence of any topic is to jot down lead points on a card and cover the subject by lecture method.

IMPORTANT NOTE: White Boards with Marker pen and Duster are available online. If convenient to carry them to different venues, you may procure and use them. In that case replace 'Chalk Board' in the text by 'White Board.

*****

5. SESSION-WISE THINGS YOU NEED

For all sessions:

a.   Ample supply of white and colour chalks and 2 quality dusters.

b.   Arrangement for tea/coffee in the recess between sessions.

Abbreviations used:

1.   Reference to page No. = P. (No.)

2.   Right Hand Column = RHC.

3.   Left Hand column = LHC.

4.   Exercise = Ex.

Day/Session	Appendixes/ Handouts,	Physical Items/Equipment	Cards Listing
1st Day **Session-I**	CPS-01/I CPS-02/I	(Copies of Suggested Time Schedule- Page 8)	Words Meaning Uneasy P20/LHC – Categories of Problems (21)
Session-II	……….	……….	– Methods of Solutions. – Points from Pp 28–31
Session-III	CPS-03/III	– Recorded Sound- Ex.14(a) – Audio Aid. Ex. 14(b) – Powder in bottle-Ex.15 – Eucalyptus in bottle- Ex.17. – Emery Sheet – Ex.18(b) – Video Clip – Ex. 19	Mental Abilities (p.40)

Day/Session	Appendixes/ Handouts,	Physical Items/Equipment	Cards Listing
Session-IV	AP-01/IV AP-02/IV	– Soap Cake or Candle-Ex.23. – Transparencies of: A-01 & A-02	Creative Abilities (p.56) – Characteristics of Creativist (P.57).
2nd Day **Session-V**	AP-03/V to AP-06/V (4 Nos.) CPS-04/v	– Drg. Of Two Concentric Squares. – A Piece of Square Pipe-P80 – Overhead Projector.	– Mental Blocks(List of sub-heads: Pp. 67–86) – Rapid Fire Questions Ex.40 (P75)
Session-VI	AP-07/VI AP-08/VI AP-09/VI	....	– Questions From Page 94. – Creative Techniques: Pp. 97–99 – Brainstorming Rules P. 101
Session-VII	....	Video recording of Brain-Storming of 1st Problem.	....
Session-VIII	AP-10/VIII	....	....
3rd day **Session-IX**	....	....	– Factors- P. 117 (Last set on LHC) – Attributes- P. 117 (LHC) – Factor x Rank – 118 (LHC) – Matrixes/Tables – Pp. 119–129 Total 12 Nos
Session-X	**Session-IX-Continued**	....	...... do ......
Session-XI	CPS-05/XI	....	– Check List – P.133 (LHC) – Questions – P. 134 (LHC) – Guidelines – P.136 (LHC)/CPS-05)
Session-XII	CPS-06/XII CPS-07/XII AP-11/XII	....	....

***

6. SUGGESTED TIME SCHEDULE

Day	09.00 to 10.30	10.30 to 10.45	10.45 to 12.15	12.15 to 13.45	13.45 to 15.00	15.00 to 15.15	15.15 to 16.30

1st Day	**Session – I**
	* Inauguration
	* Program:
	  – Objective.
	  – Philosophy.
	* Problem:
	  – Definition.
	  – Categories.
		C O F F E E B R E A K	**Session – II**
			* Solution:
			  – Types of Solns.
			  – Methods of.
			    finding
			    Solutions.
			  – Introduction
			    to CPS.
				L U N C H	**Session – III**
					* Approach
					* Creativity
					  and
					  Civilization.
					* Mental Abil-
					  ities.
					* Creative
					  Abilities.
					         Part-I
						C O F F E E B R E A K	**Session – IV**
							** Creative
							  Abilities.
							         Part-II
							
							* Characteristics
							  of a Creativist.

2nd Day	**Session – V**
	* Mental
	  Blocks.
	* Emotional
	  Blocks
	* Perceptional
	  Blocks
	* Cultural
	  Blocks.
			Session – VI
			* Problem Finding
			* Fact Finding.
			* Idea Finding
			  (Techniques).
			* Group-
			  Brainstorming
					Session – VII
					* Brainstorm-
					  ing
					  (Members
					  Problems)
					* Notes on
					  Brainstorm-
					  ing.
							Session – VIII
							* Brainstorming.
							  (Continued)
							* Case Study:
							  (Non-Magnetic
							  Flats)

3rd Day	**Session – IX**
	EVALUATION:
	* Selection of
	  Criteria.
	* Case Study
	  (Pen Holder)
	* Matrix
	  Techniques.
			Session – X
			*Planning and-
			  Development
			  – Check Lists
			  – Execution Plan
			  – Road Blocks
			*Implementation
			* Final Report
					Session – XI
					
					* Creativity
					  Test.
					
					* Guidelines
					  for
					  Preparing
					  Report.
							Session – XII
							
							* Seminar or
							  Presentation of
							  Final Report.

Trainer is advised to procure and exhibit at least two films on Creativity/Creative Problem Solving. Plenty of films are available on Internet. (See page 212)

***

3

TRAINING TECHNIQUES OF CREATIVE PROBLEM SOLVING

SESSION — I

— INTRODUCTION

— PROGRAM OBJECTIVE

— PROGRAM PHILOSOPHY

— PROBLEM: DEFINITION, CATEGORIES

SESSION – I

INTRODUCTION, PROGRAM OBJECTIVE, PROGRAM PHILOSOPHY, PROBLEM: AREAS. DEFINITION, CATEGORIES

Procedure	Statement
INTRODUCTION	
INAUGURATION by the ORGANIZER (Refer to para. 2.11*, Page. 4) Open Session. Creating informal atmosphere	Thank you Mr. - - - - - - - (Organizer of the Workshop) Greetings, and welcome to all of you. I feel privileged to share my experiences with you. I hope our deliberations will be of much relevance to our jobs and of mutual benefit in personal life, too. Success of this course depends on your contribution and free and frank participation in the discussions. Let us, therefore, be informal. This is not the usual type of training or tutorial class. Nor there is a test of anybody's any ability. We will examine your own present style of working, and action you take on the problems and then compare it with what this programme offers. Therefore, my request is: – Please participate in the deliberations with enthusiastic interest. – Respond to the leads with child-like spontaneity without inhibitions, forgetting your surroundings, status, age, etc. The scope and application of Creative Problem Solving is very vast:
PAUSE	

<table>
<tr>
<td>

<u>CHALK BOARD (C.B.)</u>

– I hear and I forget.

– I see and I remember.

– **<u>I do and I understand</u>**.

</td>
<td>

– Some Universities in USA offer 2-year full time course. (e.g. State University of New York at Buffalo).

– But with the limited time available to us, our exercises will be something like depicting the whole story of Ramayana or Mahabharata in a 3 episodes of 3 hour each.

A Chinese philosopher said:

– I HEAR AND I FORGET.

– I SEE AND I REMEMBER.

– I DO AND I UNDERSTAND.

In this programme the emphasis is, therefore, on DOING things and the workshop is well loaded with ample exercises, experiences and case studies.

HERE WE GO and EXPLORE the power of imagination.

</td>
</tr>
</table>

PROGRAMME OBJECTIVE

(MAY BE DROPPED FOR 3 – DAY PROGRAMME)

⇩　　　　　　　　　　　　　　　⇩

<table>
<tr>
<td>

<u>Exercise No. 1.</u>

Copy the questions on a card and pose them in your normal style.

</td>
<td>

Think of any one of colleagues or a participant in this group and institutionally rate his traits.

Is he:

– Self confident or timid?

– Submissive or domineering?

– Reserved or talkative?

– Open minded or close minded?

– Curious or lethargic?

– Negativistic or self starter?

(Contd.)

</td>
</tr>
</table>

Procedure	Statement
	– Intuitive or intelligent?
	– Cooperative or quarrelsome?
	– Avoids responsibility or accept it?
	– Gives or takes credit?
	– How does he act in crices? etc.
Allow 5 minutes	
	What is your overall rating?
	– Above average?
	– Average?
	– Below average?
Let 1 or 2 members spell out the rating without disclosing identity of their colleague.	Please tell us your assessment without disclosing identity of your colleague.
	You have rated him from the first impression, comparing with stereo-type.
	Still intuition played a great part.
Exercise No. 2.	Next, you will introduce to each other. The contents of introduction should be crisp and precise like a bikini.
	What does a bikini do?
Pause for 2 or 3 funny responses.	Well, any one of you should be able to introduce any other member regarding his:
	– Personal/professional background.
	– Accomplishments like patents, poems, articles, hobbies, etc.
	– How he would like us to address him during the programme.
	– What he expects from this programme.

Procedure	Statement
<u>C.B.</u> 1. Personal. 2. Accomplishments. 3. Pet name. 4. Expectations. After completion of introduction, ask 2 or 3 participants to introduce 1 or 2 other participants.	
	Please face each other and discuss among yourselves at least one of the accomplishments of your colleague in the group for about 10 minutes and jot down only 2 or 3 striking features.
Allow about 10 minutes. When the discussion is over, ask 2/3 participants to disclose their findings. <u>Sum-up (C.B.):</u>	Every individual has a window like this.

	Known to self	Not known to self
Known to others	1 OPEN	3 Blind
Not known to others	2 HIDDEN	4 Unknown

Procedure	Statement
	When you entered the room, you had exposed a very little of your 'OPEN' self, not necessarily in words but may be by facial expressions, gait, or style of talk, dress, posture, etc.
	By discussion, you have revealed some accomplishments i.e. 'your hidden self' or private personality, thereby increasing the area of 'open self, or public personality and your friends have probed and exposed some part of 'blind' sector, thereby reducing the 'hidden' and 'blind' portions.

Procedure	Statement
Shift line AC to right and BD to lower side as below.	

Point to 3rd and 4th sectors

C.B.

OBJECTIVE:

– EXPLORE THE UNKNOWN. $\longrightarrow$

– IMPROVE CREATIVE POTENTIALS. $\longrightarrow$

Objective of this programme is to explore and reduce these blind and unknown areas.

This may be the repository of unsuspected strengths.

– To expose unknown part of personality.

– To help/develop creative potentials and put it to effective use.

You are creative. But that potential is dormant:

– Covered with dust-like an art craft.

– Buried like a treasure.

This programme aims at uncovering it, – reclaiming it for you–like some people Reclaimed the precious land submerged in ocean at Nariman Point in Mumbai. The land existed before but people were not aware of its existence till some bright brains thought of reclaiming it.

OPTIONAL

Procedure	Statement
PROGRAMME PHILOSOPHY	
<u>**Exercise No. 3.**</u> <u>For additional exercises, see Annexure-01/I</u> Wait for 4 or 5 possible solutions like: – Released by one hand and caught by the 2nd, before touching the ground. – Dropped the egg on foam or water. – It was a fossil egg. – It was a plastic or wooden egg, etc.	I dropped an egg from a height of 7 feet. (or was it 70 feet?) But it did not break. How come?
	In traditional system of things, you would score very high in IQ test. But a creative mind delves much deeper beyond the walks of traditional boundary. What were the common, conscious or sub-conscious assumptions behind these solutions?
Rarely someone is likely to point out that it was 'g,' i.e. gravitational pull.	
	You have consciously or sub-consciously assumed the existence of gravitational pull and that you should create a counter balancing force in order to prevent the breakage of the egg. Am I right? This is a stereo type approach to a problem. Some of the solutions indicate the assumption that the egg was made of an unbreakable material. That was a 'clever' but not creative approach. But, unless you break this straitjacket thinking pattern, you may not find the several alternatives that exist <u>beyond your immediate surroundings.</u>

Procedure	Statement
	You know, a man can live and survive in 'zero gravity' zones in space craft.
	He can also live and work under water
	Had you only stretched your mental muscles a little farther, you could have thought of me as the Squadron Leader Rakesh Sharma in Soyuz T-7 or Captain Cousteau diving deep in Atlantic Ocean when the egg was released.
	Yet another creative way of looking at the problem is to think of the source of the egg.
	An egg is a product of reproductive system of a female – of any species.
	I might have released a bird or cat carrying that egg from 7 ft. or 70 ft.!
C.B. PROGRAMME PHILOSOPHY: – Voyage to the Frontier of Creative Thinking.	The philosophy of this programme is to become aware that we have far greater imaginative power than what we put to use, and that we can stretch our imagination beyond the conventional thinking to the FRONTIERS OF CREATIVE THINKING.
	Each animal or creature has some means of protecting itself from the predators like claws, quills, flanges, horns, powerful kick in the hind legs, ability to emit and spray obnoxious smells, camouflaging ability, etc. However, the man has the poorest kit of survival still he survived because of his power of imagination.
	It is this unique faculty of imagination that sets the man apart from all other living things.
	The thoughts once regarded as dreams are the realities today.
Distribute the handout CPS-01/I 'It Couldn't Be Done' – a poem.	

Procedure	Statement
	PROBLEM: DEFINITION, CATEGORIES

Clean chalk board and divide it into 3 columns
 	
C.B.
1. PROBLEM – 1st column
(Cover by lecture method)
	Our topic of the day is Creative Problem Solving as the title indicates, and it contains three words:
	    1. Problem
	    2. Solution
	    3. Creativity.
	Let us briefly discuss these words, one by one.
	
	In my opinion, man uses his ingenuity in creating more problems in solving a single problem and that too without any formal training or tutoring.
	
	However, when it comes to solving a problem, it calls for a lot of efforts in the form of positive and negative stroking, coxing, cajoling, training and what not—to make him aware of his own creative potentials.
	
	Let me elaborate my statement that man creates more problems in solving a single problem.
	
	Take an example or two.
	
	Example – 1

Describe 2 or 3 examples in 'matter- of- fact' way. Other examples are given in Annexure - 02/l.	While sharpening a pencil, a child cuts his finger resulting in bleeding injury. You rush him to hospital, wait for doctor, wait for injection, dressing, etc. Then the child misses school, homework; finds difficulty in catching up with the syllabus Money is spent on his treatment and conveyance. Time is spent by mother in feeding him, and so on. Look at the thing that cut the finger. It is a knife or blade, a simple device for sharpening pencil.

Procedure	Statement
Get 1 or 2 examples.	How many devices do you recall of performing this simple function and problems associated with the manufacture, marketing, care, re-sharpening and so on?

In order to solve one problem of 'making marks' on paper, a pencil was invented which lead to a host of other problems like device for sharpening it, health hazards due to chewing of lead, pollution caused by manufacture of erasers, paints, lead, etc.

Example – 2

Take another example:

Paining of Mona Lisa – a simple painting created by the artist Leonardo Vinci to please himself.
– Look at the dust kicked up in the interpretation of her smile.
– Security problems.
– Spurious copies and cheating.
– Struggle for possession/social status.
– Headaches for insurance companies, so on and so forth.

Example – 3

Automobile was invented for solving personal transportation problem. In a short span of 20 years, it became a multi-billion dollar industry supporting millions of families all over the world, at the same time creating multitudes of problems in science, engineering, medicine, manufacturing, marketing, maintaining and so on.
• Pollution due to exhaust fumes and effluents emitted in manufacture of synthetic materials are threatening the very face of human race.
* Maintenance.

Contd./20 |

Procedure	Statement
	• Black market and spurious parts. • Social status. • Balancing of family budget • Loaning or not loaning to friends • Minor children wanting to drive. • Risk to limps and lives in manufacturing and use, etc. And a variety of problems associated each of them. These are but a few examples – like a drop in the ocean. Even in the case of laws, rules, procedures and systems we get a similar picture.
PAUSE for some responses and push on without comments	Any question or comments? Let us go back to 1st word in CPS, viz. PROBLEM. Where do you find problems and why? Problems exist everywhere – wherever man exists and this is why the hierarchy in every field of human activities – be it a family or factory, hospital or army, school or society, national or international body.
C.B. (Under '1. Problem') (a) Problem Areas	Coming to problem areas and our concern, problems are nothing but absence of ideas. Conflicts lead to problems, and conflicts are created by our own perception and images of the situation. If the gap in perceptions widens, conflicts emerge. A conflict may have some international ramification like invasion of one country by another country. Or it may be related to national, social or industrial life. It may sometimes involve us and our work life Achieving a set target of production, property disputes, daughter's marriage and the like do affect us.

Procedure	Statement
Point out to column heading: 'PROBLEM'	Going back to the topic, I would like to open our discussion by asking you to narrate briefly a few problems that you might have faced recently. Let the description be very brief like the followings: – After reaching office, you found that keys are left in The house. – Foreign dignitaries on visit to factory but the camera is not working. – Your colleague residing nearby bought a large TV set. Your wife and children start nagging – a status implied.
Get 3 or 4 examples. Add 2 or 3 examples of your own. Additional examples are given in Annexure – 03/I. C.B. 1ˢᵗ column, under 'Problem' **(b) Problem Definition:**	Having considered these problem situations, can you suggest a **definition of a problem**. **What is a problem?** When a problem confronts you, what is your reaction or feeling?

Bring out:
We feel
   – Disturbed
   – Uncomfortable
   – Frustrated.
   – Uneasy.
   – Miserable
   – Angry
   – Irritated
   – Anguished
   – Jealous, etc.
	So, **anything that makes us uneasy is a problem**.
	
	Anything that is caused by changes, interruptions & failures and which breaks work rhythm, is a problem

Procedure	Statement
	Problem implies disorder, a mess—about which we want to do something to restore "order."
C.B. 1st column, under 'Problem' (c) Categories of Problem:	
	There are various methods of classifying problems. We have already seen that problems may be international, national, social, industrial, personal, etc.
Distribute Handout: CPS-02/I. {Problem sheets}	Problems may also be classified as human relations, scientific and technical.
	Keeping in view the limited objective of our course and short time available at our disposal, we will restrict our discussion to the problems related to our jobs and/or work life, which can be solved by an innovative approach.
Make it very brief (or drop) C.B. Under 'Categories of problems' write: 1. 'Sense' type of problems (or 'smell' type)	Numerous though the problems and their classifications can be, I prefer to put them in 5 categories:
	Example – 1.
	There are occasions when we notice some change in attitude of an employee towards the others. He may become quarrelsome. Isn't it a problem?
	It is a kind of problem you 'smell.' You don't know the consequences nor can you predict how it will affect you and your responsibilities.
	Example – 2.
	A house wife dropping or throwing utensils comes under this category.
Get 1 or 2 Examples from the participants	

Procedure	Statement
	The second category of problem is called 'Anticipate' type.
C.B. Under 'Categories of problems' write: 2. 'Anticipate' type of problem (or Size-in-advance' type)	
	Example – 3.
	Occasional misfiring of an automobile engine on the highway is the indication of on-coming bigger problem.
	Example – 4.
	Your management desires to change the profit sharing bonus scheme by delinking it from annual profit and linking the payments with monthly performance.
	Many a time we get such advance notice of change or interruption which you feel will not be well received. You as a part of management or head of family, have a chance to take some preventive action.
	That is a problem you could 'size-in-advance' or anticipate.
	You can visualize its consequences and anticipate gravity of the problem. You have some chance to take action to prevent that situation developing into a problem.
Let 1 or 2 participants narrate examples of 'anticipate' type. C.B. Under 'Categories of problems' write: 3. 'Impose' type of problems (or 'Come-to-you' type.	Let us see what is the third category called 'Come to you.'

Procedure	Statement
	Example – 5. Employees sometimes ask for change in job or shift. Example – 6. Your neighbour requests you to look after his pet dog when he goes on holidays with family. When somebody approaches you with a grievance or request for some favour or help, wants you to do something out of turn, you feel uneasy. Such problems 'come to you,' i.e. they are 'imposed' on you. You require some time to brood over them. You cannot take any snap decision on such problems. Even to say 'no,' makes you uncomfortable.
C.B. Under 'Categories of problems' write: 4. 'Invited' type of problems (or 'Run into' type)	If you told your driver to wash the car and he refuses, that would be an 'invited' problem. Example – 7. Meddling with equipment like iron press or TV set without knowing anything about electrical, may result in a crude shock. By your own action or behavior, you get involved into a problem of this type. Example – 8. After removing appendicitis, surgeon forgot to remove scalpel from the stomach of patient after operation.
Invite 2 or 3 examples from the participants, but ensure brevity.	

Procedure	Statement
<u>C.B.</u> Under 'Categories of problems' write:	There is one more category called 'Search' type
5. <u>'Search the truth' type of problems</u>	
	This category belongs to scientists, inventors, philosophers, and thinkers.
	Scientists start a blind hunt – out of curiosity and intense desire to accomplish something unique – to challenge old ideas, or concept – to answer some baffling observations.
	Millions of discoveries and inventions are the product of such spirit of inquiry and courage.
	Scientists and inventors encountered many, many agonizing situations, faced scorn and neglect, endured prison & exile. And some even died at the stake – all in fulfillment of their dreams and for the sake of their ideas and visions.
	Their dreams have now become realities and their problems our comforts.
<u>Invite 2 or 3 examples of this type.</u>	
<u>TEA/COFFEE BREAK.</u>	

4

TRAINING TECHNIQUES OF CREATIVE PROBLEM SOLVING

SESSION — II

SOLUTION:

— TYPES

— METHODS

SESSION – II

Procedure	Statement
SOLUTION: TYPES & METHODS	

Procedure	Statement
<u>**C.B.**</u> Heading of 2nd column: **1. <u>SOLUTION</u>**	We have defined a problem as some 'messy' situation that causes disturbance or uneasiness. Obviously solution of problem means: – Removal of uneasiness. – Restoring order in disordered situation. Solving the problem means finding ways and means: – To clear the 'mess.' – To satisfy the problem.
While carrying out the following exercises, use spur or prompt questions to bring out 3 types of solutions, viz.: a. Removal of causes. b. Help to live with it. c. Convert problem into opportunity.	
<u>**Exercise No. 4.**</u>	While going for an early morning walk, you noticed that a cast iron cover of a man-hole located in the middle of the road is missing, Think of all possible action that you would take.
Get 4 or 5 solutions and record them in 2nd column under 'Solution': – Cover it with a bolder. – Cordon off with stones or rope. – Put on a red flag/lantern. – Inform municipal office. – Forget about it, etc.	

Procedure	Statement
Note additional ideas in 2nd column on C.B. – Use RCC cover. – Use Stone cover. – Why not FRP cover? etc.	 Let us think of the causes for the unauthorized removal of the man-hole cover. – It is cast iron and can be melted. – It can be sold to a scrap dealer. – Poverty has driven someone to commit theft. – Even if it is replaced, theft may recur. What other ideas come to your mind? At the higher level of thinking, you may ask as to why the man-hole is in the middle of the road. And at still higher level of creativity, we may question the very existence of man-hole itself. Use of compressed air or water and/or slurry pump may automatically come to your mind as an alternative solution.
C.B. in 2nd column: Convert man-hole into Hose Hole.	Type of solutions that were suggested by you, fall under 3 types, namely: – Removal of causes of theft i.e. replacing CI cover with Concrete or FRP cover. – Convert problem into opportunity. i.e. use of cheaper Material like RCC. – Helping to live with the problem – what cannot be cured should be endured. Go home and forget about it.
C.B. Under SOLUTION: **TYPES OF SOLUTIONS:** 1. Removal of causes. 2. Converting problem into Opportunity. 3. Living with it.	

Procedure	Statement
(Other suggested exercises in place of or in addition to the above exercise are: – Roof or Water tank/drum Is leaking. – Head lamp of a vehicle is fused. – Float of a float valve is damaged.) **METHODS OF FINDING SOLUTIONS:** Cover by lecture method. <u>C.B.</u> **Methods:**	 The methods of solving problems depend on type of problems and the type of need that has to be fulfilled at a given point of time. For example, we have routine problems and routine solutions. – Felling hungry? – eat food. – Missed the bus? – Take a taxi. – Hinges are squeaking? – Put a drop of oil. – Dog is chasing you – run for life.
<u>Additional problems of routine type are given in Annexure – 04/II.</u> Add more from psycho-cybernetics. <u>C.B.</u> (Under 'Methods') 1. <u>Routine problems-</u> <u>Routine solutions.</u>	People develop a number of auto mechanism for solving routine problems Habit is one such mechanism which is most commonly and almost automatically used in solving routine problems. Let us consider another set of questions. – Shall I invest in stock (Shares)? – Should we allow girls to date? – Should we keep a dog as a pet? – I found a Rs. 10 note – shall I keep it or -----?

Procedure	Statement
For additional problems of 'Judgmental' or Logical type, see Annexure – 05/II.	Such problems call for decisions between: "yes' or 'no'; 'right or wrong'; 'true' or 'false'; 'moral' or 'immoral,' etc.
	Such problems are grouped together as 'judgmental' or 'logical.'
C.B. (Under Methods): 2. Judgmental problems and Judgmental solutions.	
	The 3rd category is called 'Analytical problems and analytical solutions.'
Exercise – No. 5. (Farmer's story)	A farmer wanting to cross a bridge has a tiger, a goat and a bundle of grass. The rules permit only one item to be carried across the bridge with a person.
	Help the farmer in crossing over the bridge with his possessions in least time.
Allow 5 to 10 minutes and ask for number of 'to and from" trips.	
Copy the following chart on a card and reproduce on C.B.	
Start: Tiger + Goat + grass Step-1: Tiger + grass → goat. ← farmer Step-2. Grass → tiger ← goat Step-3 Goat → Tiger + grass ← farmer Step-4 → Goat + tiger + grass	
	Such problems and riddles are for exercising the mind muscles. **Logic is suspended**. No question is asked as to why the farmer wants to rear a tiger.
Exercise – No. 6. C.B. 91, 82, 73, ---, 55, 46, ---?	What are the 2 missing numbers in the series: 91, 82, 73, ---, 55, 46, ---?
Allow a minute or 2.	

Procedure	Statement
Exercise – 7 C.B. Draw the diagram: Furnish the solution, if required Hint: Divide ABCD into 16 small squares – 4 column and 4 rows. Shaded square contains 4 small Sqs. Additional exercises are given in Annexure – 06/II. Make photo copies of these exercises and give them to the participants as 'Homework' for them and their families. (NB: Answers to problems are given in Annexure – 06-A/II C.B. under 'Methods': 3. Analytical problems – Analytical solutions.	Divide the un-shaded area into 4 equal parts of the same shape. Shaded area is exactly ¼th of total area. Mathematical and pure research problems fall under this, i.e. analytical category. Such problems call for heuristic (step by step) method in solving them and yield only one correct solution.

Procedure	Statement
	Fourth method is the theme of our workshop.
	Creative method of solving problem needs, in the first place, creative problems.
Exercise No. 8.	In what ways can you stop rusting of drinking water pipe lines?
Bring out 4 or 5 methods, including replacement of G.I. pipes by plastic pipes	
Exercise No. 9.	What other words will convey the same meaning as the word 'stifle' in the following sentence:
	Negative attitude 'stifles' creativity.
Wait for 8 or 10 new words.	A group of High School students thought of more than 25 words in 10 to 12 minutes, like:
	Cramps, jams, curbs, clips, strangulate, smothers, snubs. dampens, throttles, kills, curtails, stamps, dulls, crushes, paralyses, cripples, crucifies, mortifies, murders, massacres, invalids, pulverizes, chastens, crooks, destroys, petrifies, beheads, debilitates, etc.
For additional exercises on 'Creative Problems.' refer to Annexure - o7/II.	
C.B. (Under 'Methods'): 4. Creative problems – Creative solutions	
	Creative types of problems have a very large number of solutions, none of which is absolutely correct but most of them are workable, now or later.
Creative Methods:	
C.B.– Third column 3. Creative	Looking at a problem from unusual angle, using fantasy and analogies, brainstorming, playful combinations, etc. are the main features of creative methods.

Procedure	Statement
<u>Exercise No. 10.</u>	Let us carry out one more exercise.
	What are the functions or uses of a 'Sari'?
<u>**C.B.**</u> Get 4 or 5 uses, record on C.B. and read out the rest. – Cover body. – Attract attention – Please oneself. – Please husband. – Enhance prestige. – Enhance beauty. – Spite neighbor. (Humiliate neighbor) – Maintain domestic peace/bliss. – Indicate festivity. – Exhibit affluence.	
	These answers are based on traditional wisdom.

Read slowly with pauses between two questions	The list would have been much longer, had we asked of ourselves questions like:
	
	– What is its texture?
	– Is it porous?
	– Is it coarse?
	– Is it smooth? – Flexible?
	        – Absorbent?
	– Can it be cut? – Stitched?
	      – twisted?
	      – squeezed?
	      – folded?
	      – coloured?
	      – shredded?
	      – burnt?

Record additional uses on C.B.	Well, now can you think of some more uses of sari? I hope the meaning of 'creative' is clear from this and previous 2 or 3 exercises.

Procedure	Statement
Creative Problem Solving (CPS)	
INTRODUCTION TO CPS	By "Creative" we mean: innovative, having an element of novelty and usefulness.
	Our discussion so far on the 3 words (problem, solution and creative) should enable us now to define the process called "Creative Problem Solving ".
Definition of CPS.	**Creative problem solving simply means handling a problem in an imaginative, innovative way, – in a new, better way, which results in multiplicity of useful alternatives.**
C.B.	
– Imaginative way.	
– Innovative way.	
– A new, better method.	Before deliberating details of CPS, we need to understand different Mental and Creative Abilities.
	These abilities will be taken in the next session.
Assorted Types of Exercises:	While carrying out the following exercises, please indicate their 'TYPE,' wherever possible.
Announce answers/solutions after receiving response to each 'problem.'	
Answer/Type	
1. 6 (Analytical)	1. If 4 loaders (Hamals) take 3 hours to load a lorry, how many loaders are required to load the same lorry in 2 hours?
2. Yes or No. (Judgmental)	2. Should you give a smart phone to your son as a gift for passing out High School in 1st division?
3. Many Ways (Creative)	3. In a commonly used tap (faucet), threads are 'slipped.' In how many ways can you stop wastage of water?
4. Yes. Many ways. One is to remove ball holder & suck in ink. (Creative).	4. Can you refill an empty barrel of ball pen with pen ink? If yes, how?

Procedure	Statement
Answers:	
5. Many ways. (creative). Simplest way is: Put a tray full of water on window sill.	5. One of 2 shutters of a window in my room is always kept open. The window is fitted with grill of horizontal bars 4.5"apart. How to prevent a stray cat from passing through the window?
6. All (but response will be Feb.)	6. Which English month has 28 days?
7. Cylindrical or spherical shutter pivoted along its axis.	7. A door has a single shutter pivoted at 2 points in the door frame. It is capable of rotating but even if it is rotated through 360°, nobody can pass through the door. How come?
8. Incorrectly.	8. Which word can't be spelt correctly?
9. Hang hammock-like foldable bed between berths.(Good for children).	9. How to increase capacity of a sleeper coach?
--	--
Note on Riddles, Puzzles, Cross-words, SU-DO-KU.	Before closing this session, let me add that solving Riddles, Puzzles and Crosswords provide great opportunity in strengthening mental muscles but it should not become obsession (to kill time).
	Do spare some time in solving such problems for sharpening your creative abilities.
Read out from Annexure-15/II.	Here are some Exercises for you.
LUNCH BREAK.	

5

TRAINING TECHNIQUES OF CREATIVE PROBLEM SOLVING

SESSION — III

MENTAL ABILITIES & CREATIVE ABILITIES

CREATIVE PROBLEM SOLVING

(A Voyage to Frontiers of Creativity)

SESSION — III

MENTAL ABILITIES AND CREATIVE ABILITIES

Procedure	Statement
C.B. **– MENTAL ABILITIES** **– CREATIVE ABILITIES – Part – I**	
⬇ (OPTIONAL FOR 3-DAY COURSE) ⬇	
APPROACH **C.B.** **Exercise-No. 11.** (Meditation) 	Creative process begins with identification of problem for creative attack. And then goes through the recognition of creative abilities, mental blocks, strategies for unblocking, various steps of CPS, namely: fact finding, problem finding, idea finding, etc. ultimately ending in effective action. In covering these aspects, I may not follow a traditional, sequential, or relay race approach like this: Point to top Fig. on LH Column. Nor will it be purely PERT/CPM approach. Or a holistic i.e. rugby approach like this: Point to lower Fig. on LH column. The approach will be a combination of sequential and holistic methods, not because of my liking or disliking of any particular approach but because the very nature of the subject (CPS) is like that. -- Let us do a small exercise on meditation—not for seeking salvation but for gaining some insight – in the state of mind. – Please check which nostril is breathing

Procedure	Statement
	Please note that both nostrils are not breathing at a time.
	How many of you were aware of this?
	When the left nostril is breathing, the right side of the brain is more alert and vice versa.
	Please sit in relaxed posture.
Give long pauses between instructions.	
	Close your eyes and concentrate on breathing.
	Imagine a square.
	Hold it.
	Introduce a triangle of the same height next to the square.
	Hold them.
	Finding difficult to focus attention of the figures?
	Try.
	Concentrate.
	Now, change the places of the square and the triangle.
	Hold them.
	Now imagine a circle of the same diameter as the height of the triangle, placed by the side of the square.
	Hold all the 3 figures in a line.
	Now, introduce colours.
	– Circle is red.
	– Triangle is yellow.
	– Square is blue.
	Hold the three figures in a line–triangle, square. Circle.
	Interchange places of circle and square.
Pause.	Hold them.
Announce completion of the exercise	

Procedure	Statement
	You can now open your eyes and resume normal posture.
	Please produce the 3 figures with colours in their last sequence.
Disclose the sequence on **C.B.:** Y R B Wait for responses.	
	How many of you got it correct?
	Those who could hold both the shapes and the colors are emotionally well balanced, say the psychologists.
	And that is an important perquisite in creative problem solving.

---------------------⇑ **(O P T I O N A L)** ⇑--------------------

Procedure	Statement
	Creativity And Civilization
C.B.	Power of imagination is limitless.
	Ideas can make or unmake a society or a nation.
Creativity and Civilization	
Cover by lecture method	Economic prosperity of a society or country rests upon the creative abilities of its people rather than upon its natural resources.
	Poorest people of our country are probably more in state of Bihar where the richest deposits of mineral wealth are found.
	Japan has no minerals of its own, but it is one of the most (industrially) advanced countries.
	Think of progress made by man in any field.
	For example, a single idea of using fire has generated the whole world of automotive industry, which is providing employment to millions of people all over the world.

Procedure	Statement
	Petroleum Industry, metallurgical industry, plastic and paint industries are providing employment to additional billions of people.
	Just imagine the scene of a man using a hand fan and think of the millions of ideas that have gone in making the automatic air conditioning of sky scraper buildings.
Exercise No. 12. Participants to narrate progress in 2 or 3 fields.	Think of various stages of evaluation, and draw a brief word picture relating to: – Transportation system. (Stone wheels, animal driven carts, automobiles, boats, ships, airplanes—to spaceships.) – Communication (Runner mail to satellite communication.) – Food processing (Firewood to microwave ovens.)
Additional topics: – Entertainment. – Medicines. – Aircrafts – Construction. – Mining. – Agriculture.	All these developments speak volumes of the value of imagination. Such creative imagination was coined as '**imagineering'** by the Aluminum Co. of America. It means: 'let your imagination soar sky high and wide, and engineer it down the earth.' Osborn, the father of Brainstorming Techniques concludes that history of civilization is essentially a record of man's creative abilities.

Procedure	Statement
MENTAL ABILITIES	
Creativity and Intelligence	Creativity is one of the most important components of intelligence, which in itself is a marvelous composition of: – Reasoning ability; – Skill in solving problems; – Personality; – Structure and chemistry of the brain; – Creativity, etc.
C.B. **MENTAL ABILITIES:** Complete the exercises as quickly as possible:	Creativity, i.e., power of imagination, in turn, encompasses many mental abilities.
1. Exercises on Absorptive, Retentive and Recall abilities.	Let us carry out a few exercises on the topic i.e. Mental abilities. The first group of exercises is on Absorptive, Retentive and Recall Abilities.
Exercise No. 13(a): Draw simple figures of 1 or 2 objects on chalk board, say, a leaf, flower, tumbler, hut and/or a bird, and ask the participants to name it/them.	
Exercise No, 13 (b): Ask the participants to complete 1 or 2 sums like these: $$4 + 4 = ? \qquad 9 - 4 = ?$$	
Exercise No. 14 (a): Participants to listen to 'noises' coming from outside and to recognize their sources. (Men talking, birds chirping, dragging, vehicle moving, dog barking, tinkering, etc.)	

Procedure	Statement

Exercise No. 14 (b):
 	
 A still better exercise is to play a recorded sound (from your mobile phone or tape recorder) and ask the participants to recognize it.
 	
 e.g.   – A few lines of a popular song or its tune.
        – Crying of a baby.
        – Starting of a vehicle.
        – Mewing of a cat, etc.
 	
 Exercise No. 15:
 	
 Circulate a bottle containing white powder (sugar or glucose) and ask the participants to identify the substance.
        Announce: ⟶
 Note the tests they carry out, e.g. 'seeing,' rubbing between fingers, smelling and finally tasting.
 	
 Exercise No. 16:
 	
 Ask 1 or 2 participants to narrate briefly some funny (or sad) incident/episode.
 	
 Exercise No. 17:
 	
 Pass on a bottle containing eucalyptus oil or asafetida powder and ask them to identify it.
 	
 Exercise No. 18(a):
 	
 Participants to narrate the feeling they get after touching the drawer and plated handle or/and underside and topside of a table.
 	
 Exercise No. 18(b):
 	
 A still better exercise is to circulate a sheet of 'water paper' (600 grit emery sheet) and ask the type of sensation the get by touching the coated and plain sides.
 	
 For additional exercises, refer to Annexure - **08/III.**
	
	 It is edible
	
	 It should be a real life experience.

Procedure	Statement
2. Exercise on "Reasoning or Judgmental Abilities" Additional exercises in Annexure - 09/III.	Our next exercise is on 'Reasoning or Judgmental 'Abilities.'
Exercise No. 19:	
Participants are given one of the topics noted in the 'Statement' column or any other topic of current relevance for discussion to bring out the reasoning/ judgmental ability.	
The topic should be somewhat controversial but not hurting	What are your views on (any one): – State of affairs in Sports area? – Quality of TV programmes? – Language policy of Central Govt.? – Salaries/perks of Lawmakers?
It should be impressed on the members that they will have to reach definite conclusion which would be recorded.	
Allow about 15 minutes.	
If possible, video-record the discussions to play back while summarizing.	
Note striking features of discussion like arguments, cross talks, somebody monopolizing discussion, someone keeping aloof, etc.	
3. Exercises on "Creative Abilities": (For additional exercises, see Annexure - 10/III.)	Let us now discuss "creative ability."
Exercise No. 20: Any one item. Allow about 10 minutes. Add 2 or 3 uses of your own.	List as many new, interesting and unusual uses of: (a) Rolling pin or (b) brick.

Procedure	Statement
Allow about 10 minutes. Add 2 or 3 uses of your own. **Exercise No. 21:** **Allow 5 minutes and push on.** **Exercise No. 22:** Additional exercises of this type are given in Annexure - 11/III: (FOOD FOR THOUGHT)	 Think of all possible excuses for reaching late to the office. Dr, Modi and Dr, Billmoria had a roaring practice in Mumbai and 'whale if time' enjoying 5-stat comforts. On one week-end day they were driving down from Mumbai to Pune when they sighted a lonely lady struggling with her broken down car in Khandala Ghat. Dr. Billmoria slowed down and when they were a few Metres away from the lady in distress, Dr, Modi impulsively exclaimed "Oh! It is my wife." On hearing this, Dr. Billmoria pulled out a revolver and shot Dr. Modi dead.
Get 4 or 5 responses.	Think of all possible reasons for this extreme action of Dr. Billmoria.
Push on without commenting at this stage.	(To the participants: Please do not disclose the answer at this stage, if you have heard this story earlier).
	That gives us enough material to sum up different abilities of our brain. ----------------------- Psychologists tell us that our brain has four fundamental areas or mental abilities
SUMMARY: **C.B. (Under Mental Abilities)** 1. Absorptive Ability:	 – First one is called 'Absorptive' ability.

Procedure	Statement
	– First one is called 'Absorptive' ability. This ability enables us to receive information and sensations from outside the world through the five channels, namely: eyes, ears, nose, Tongue and touch. ("Panchendriyas")
C.B.: (Under Mental Abilities): 2. **Retentive Ability.**	The absorptive ability to observe and focus attention is closely associated with the second ability called 'retentive power.' ------------------------------- Ability to memorize and to recall is the storage area in brain that files knowledge, information, sensation, and from which both the facts and emotions can be recalled. Referring to our exercises on identification of figure(leaf or hut), sums (2 + 2), naming of article, etc. you could identify, figures, sound, and smell because of your ability to recall the stored information and sensations. When you looked at the picture on the chalk board, the image entered the 'reception' hall of the brain.
*Name(s) of the object(s) should match with the figure(s) drawn on chalk board under Ex. No. 13(a). To stress the word "matched," you may dramatize it by holding your hands in 'namaste' pose, raising them high and rotating.	The 'receptionist' analyzed the information as a photograph or a picture, scanned the trillions of indexed photo files, compared the 'visitor' with the record and when it 'matched' with the outline of the recorded image, 'declared' it as a *(leaf or hut). The processing took place in trillionth of a second. But it did take time and it did take place. Similar was the process of identifying powder, sound, smell– Scanning through the colours and other sensation files.

Procedure	Statement
*As applicable.	When the taste or sound matched with the one that was recorded earlier, you have identified as *(sugar or salt) and the sound from *(source of sound). All types of exams., quizzes, etc. rely heavily on this ability. --- The 3rd area is termed 'Reasoning Ability.'
C.B. (Under 'Mental Abilities'): 3. **Reasoning Ability**. (Note: Play back the video strip if the discussion was recorded.)	"Ability to analyze, to synthesize, to compare, to choose, to judge," as Osborn put it, was demonstrated by your debate on -------(subject of discussion in Ex. No.19.) Normally, the discussions take the form of arguments and counter arguments. Each member tries to convince all others that his idea is better or more logical than others. Cross talks, raised voices, wide gestures and derisive laughter pre-dominate such meetings. Your discussions were no exception.
Point out a striking features of discussion held under Ex. No. 19.	No sooner an idea is thrown up than the guns are trained at it to shoot it down and the idea owner is provoked to defend his brain child with all his wit and might. The spark of offensive and defensive verbal war is thus ignited– discussion ending in a decision 'to meet again.'
Continue by lecture method. You may condense the lecture by covering the main points	Discussions in meetings of Co. Boards, Commissions and even in Parliaments are no way better or different from what we have witnessed few minutes ago. This judgmental area evaluates the new information and helps us in deciding whether it is good or bad, true or false, right or wrong, moral or immoral, superior or inferior, Legal or illegal, etc.

Procedure	Statement
	But most of the time most of us use too much of judgmental ability which dampens the spirit of enquiry instinctively to score over the others—casualty being new ideas.
	We have been trained right from childhood in this art of comparison and evaluation.
	Topics like 'pen vs sword, city life vs village life, etc. in schools, questionnaire in IQ tests, objective type of question papers in competitive examinations, and so on provide fodder for sharpening our decision making ability.
	The parents compare and actually condemn each other's relatives in presence of their children as regards their social status, poverty of manners, money power and what not.
	Then we apply our yardstick to the behavior of our neighbours, friends, teachers, and so on.
	As small children we imagined and played with pebbles, bottle caps as coins or wheels, broom as a horse or bat, little sister and brother as Mom and Dad.
	But in the environment of home and school that is created by adults, we as children soon learn that there are right and wrong things, true answers and false answers, good manners and bad manners, so much so that the curiosity, which is the corner stone of creativity, is crippled and the child is soon cast in the mould of stereo type mediocre.
	But rarely do we care to develop the art of appreciating creative talent'
	Let me make it clear that by the foregoing comments, I do not mean judgmental ability is the killer of creativity. On the contrary, evaluation i.e. decision making skill, is as important as the brakes on a car or landing gears on an aero plane.

Procedure	Statement
	We cannot do away with them but at the same time we should know how and when to use them judiciously.

	The 4th functional part of our brain is the creative ability.
C.B. (Under Mental Abilities) 4. **Creative Ability.** * As applicable. ** If the story is from Annexue-11, substitute applicable names.)	Ability to process and combine bits of knowledge and experiences to form a new pattern in a novel way is amply demonstrated by our exercise on new and unusual uses of *(Rolling pin/brick) and multitude of solutions given by you. Coming to story of **(Dr. Modi and Dr. Billmoria), ever since you have received the story inputs your conscious and unconscious mind is working vigorously working on it.
	The brain is churning out various available elements of sensation and experiences, generating new perceptions, tossing them up and down, shoving them here and there, sorting and combining the bits—all in apparent bid to form recognizable pattern, plausible explanation, that is- solution.
	Here we are speculating, visualizing and generating ideas.
* Rephrase the question to suit the story. Get a few ideas and disclosed the solution.	* By the by, what are your ideas on Dr. Billmoria's action?
	Dr. Billmoria was the girl friend of Dr. Modi who had kept her in darkness about his marriage.

Procedure	Statement
Summary:	Summing up of our discussions, our mental abilities are:
Point to heading on C.B.	1. **Absorptive Ability** to observe and focus attention.
	2. **Retentive Ability** to memorize and to recall.
	3. **Reasoning Ability** to analyze, to compare and to choose/ judge.
	4. **Creative Ability** to visualize, to speculate (to foresee) and to generate ideas.
	First three functions of the mind are being performed very efficiently by the modern machine called computer.
	But no machine will ever be capable of taking over the last function, namely creativity, from the man-kind.
*(or Ms. Rita and Mr. Masta Ram as appropriate).	In the story of Dr. Modi*, we have experienced the feeling of interplay between various mental abilities in solving a problem.
	The story has also demonstrated how our thinking process gets conditioned to follow a beaten path and how it refuses to escape the 'straitjacket.'
Distribute Handout No. CPS-03/III. (Poem- 'Calf Path')	
Allow 7–10 minutes to read the Poem.	
TEA BREEAK	

6

TRAINING TECHNIQUES OF CREATIVE PROBLEM SOLVING

SESSION — IV

CREATIVE ABILITIES — Part — II

SESSION – IV
CREATIVE ABILITIES – PART – II

Procedure	Statement
CREATIVE ABILITIES:	So far we discussed mental abilities, creative ability being one of them.
	Creative ability is not a single ability but a bunch of many interrelated attributes.
1. Ability to View Familiar in a Unique Way.	Let us see what they are.
Exercise No. 23.	
Show a cake of soap or a candle or a chalk. **(Any one)**	Think of all objects that come to your mind by looking at this object.
Record items on board.	
Ask a few spur questions.	Is it hard?—Soft? –Porous?– Brittle? Can it be cut?—Carved?—Shaped?—coloured?
Record additional ideas.	
Curiosity.	**Example:**
(Story of a sculptor).	A school boy used to peep into a sculptor's studio on his way to and from his school.
	One day when the sculptor had finished his work and looking at it admiringly, the boy approached the sculptor and enquired," Uncle, how did you know that this beautiful Lady was inside that bolder on which you started working a few months ago?"
	It is needless to say that the sculptor had a sort of x-ray vision which could penetrate deep inside a solid rock to bring out that "beautiful lady" from it.

Procedure	Statement
* As applicable.	The ideas that were spelt out by you by looking at – *(soap/candle/chalk) are the result of such inner vision.
	That is, ability to view a familiar object in a unique way –an unfamiliar way
C.B.	
1. Familiar to Unfamiliar/Unique.	
--	
2. **Ability to See Multiple Things in a Single Object.**	
Exercise No. 24.	
Draw a figure of triangle, T or circle on C.B. (Any one).	
Δ OR T OR O	What comes to your mind by looking at this figure?
– Give 2 or 3 examples of your own ideas.	Please remember: question is not what this symbol stands for, but what ideas and feelings come to your mind.
– Note ideas on the chalk board.	
– Use prompt questions.	
– Encourage combination of ideas.	
– Record additional ideas.	
	Well, this is what we call '**divergent thinking.**'
2. Multiple of things in a single object or Divergent thinking.	
--	--
3. **Ability to Interpret Unfamiliar in Imaginative Way**.	
Exercise No. 25.	
Draw <u>one</u> of the following figures on CB.	
	What comes to your mind by looking at this picture

Procedure	Statement
Get 2 or 3 interpretations.	
Exercise No. 26.	What is white, soft and edible?
Answers are likely to vary from 'idli,' banana, butter to mushrooms.	
Get 5 or 6 answers.	
	The type of life one is supposed to live after death in Heaven or hell visa-vis heavenly rewards and hellish punishments for the earthly deeds, and also, the concept of God are excellent examples of imaginative interpretation of unknown.
C.B. 3. Imaginative Interpretation of Unfamiliar.	
---	---
4. Ability to Make Transformation (Modification)	The most widely used creative ability lies in transformation and modification.
Exercise No. 27.	If the numbers 1, 2, 3, etc. stand for A, B, C, etc. in that order, what does the number 9451 represent?
Pause for obvious response 'IDEA.'	Here we have transferred numbers into alphabets to form a word.
	Many a time you might have seen children playing mathematical games with number plates of vehicles.
	They add the numbers, subtract them, invert them, convert alphabets into numbers and so on.
	This is another example of modification of available information into something significant and recognizable.

Procedure	Statement
Exercise No. 28(a). **C.B.** $2H_2O + O_2 =$ **Exercise No. 28(b).** $a^2 + b^2 = ?$ a \ c b Exhibit: Appendix: AP-01/*IV* (J bolt) OR Appendix: AP-O2/*IV* (Bird) OR Any other photograph. (The better method is to make a transparency for projection on large screen)	$2H_2O + O_2 = ?$ In these 2 examples, we have transferred some phenomena, some facts into symbols and notations. Development of chemistry, mathematics, physics, in fact, for that matter, advancement of all branches of engineering and technology were made possible by the invention of symbols and notations. A drawing or photograph is a method of transforming some relationship from its original setting into another setting where the relationship can be studied more conveniently. Please remember that drawings, photographs and charts are nothing but different shades of colours on a flat paper possessing none of the properties of the original 3-D object, and they are very useful in understanding the relationship between the various elements of original object. **Example – 8:** Thermometer is a method of transforming temperature into a length of mercury column related to graduation on its glass stem.

Procedure	Statement
	Example – 9 Watch is a similar example of transforming time into position of 2 or 3 metal pieces (called hands) related to another metal piece called dial. In fact, all measuring instruments, calculators including computers, are just methods of transforming some phenomena into something else that is easy to study and understand. Once the transformation is made then the relationships within drawings or models themselves indicate how the system will work and suggest possible direction for innovation.
C.B. 4. <u>Ability to make Transformation.</u>	**This is called Ability to make transformation.**
5. <u>Ability to Synthesize Isolated Systems in New and Original Ways.</u>	After a while, you will be given 2 stimulus words. You must combine them in an ingenious way to produce some interesting word picture. For example, I would combine 'Hairy Skin' and 'Rotten smell' to think of: – Cater-pillars. – Beautiful butterflies. – Enchanting garden. – Small children with divine innocence – rolling on lush green grass. – Locomotion without wheels. – Skill in negotiating curves. – Flexible wire brush. – Flexible shaft for wire brush powered by motor which can clean 'U' – tubes of Heat Exchangers.

Procedure	Statement
Exercise No. 29. One or two participants to give word picture.	What are the things associated with 'blue colour' and 'loud noise'? (Or 'sweet smell' and 'delicate structure'?).
Exercise No. 30. **C.B.** FACT, FIVE, FATE, VICE, EFFECT, ACTIVE AND CAT. Allow 2 or 3 minutes and then disclose the answer: 'AFFECTIVE.'	Can you identify a single word from which these words are derived? When we face a problem, our mind start playing with scores of ideas not only related to present problem but also to the previous experiences (which may not be related to the present situation) and tries to synthesize them into a solution.
C.B. 6. <u>Ability to Synthesize Isolated Schemes in New and Original Ways.</u> -------------------------------	Ability to combine the ideas from seemingly unrelated fields into a combination that can be related to the problem on hand, constitutes yet another dimension of creative ability. --
7. Ability to Fantasize:	You might have noticed small children playing 'adult' games. They transfer themselves into roles of Mom and Dad, police and thief, teachers, doctors and pirates. They collect all sorts of odd things and treat them as tools of their imaginary profession. In fact, children are keen listeners and observers. They grab on to the simplest of suggestions and turn them into fanciful ideas in their head. Part of such precious wonderment comes from a whole-hearted belief that the fantasy unfolding before their eyes is real.
Additional Exercises are given in Annexure-12/IV	

Procedure	Statement
C.B. **8. Ability to Think in Terms Metaphors, Similes and Analogies.**	– Children spilled all over the ground like cotton balls. – Butterflies are fluttering in stomach. – Bus discharging and inhaling passengers. – Lighting welding cracks in the sky – Life is like a rainbow, it needs both downpour and Sunshine to create colours.
--- **Summary**	--- We can sum up our discussion in the form of a definition of creativity. Creativity is ability: – To view familiar in unique way. – To see multiple of things in a single object. – To synthesize isolated schemes. – To transform. – To modify. – To fantasize. – To think in terms of metaphors, similes and analogies.
C.B. Creativity Familiar Unfamiliar (Known) (unknown) Learning	

Procedure	Statement
CHARACTERISTICS OF A 'CREATIVIST'	
<u>**Lecture method**</u>	Behaviour of creative people is characterized by their:
	– Ideational fluency,
	– Flexibility,
	– Originality,
<u>**C.B.**</u>	– Elaboration Ability and so on.
– **Ideational Fluency.**	
– **Idea Flexibility.**	
– **Originality.**	
– **Elaboration.**	
1. <u>**Ideational Fluency.**</u>	If I ask you to list uses of a tooth brush, you may come up with 5 or 10 or 20 ideas.
	The greater the number of ideas or solutions generated
	in response to a given problem, the greater is your <u>ideation fluency.</u>
	However, a mere large stock of ideas is not the true index of creative idea fluency.
	For example, if the uses of tooth brush include:
	– Brushing of teeth,
	– cleaning of ornaments,
	– cleaning of bicycle parts,
	– polishing shoes, etc.,
	The ideas may be many, but they are similar—all revolving round a single function, namely, dislodging dust or dirt.
	On the other hand, if the ideation leads to generation of a large variety of ideas in many categories outside the usual field of application, then you have a very important trait called **idea fluency**

Procedure	Statement
Continue by lecture method. 2. **Idea Flexibility.**	**Idea flexibility.** Consider properties of materials of which the tooth brush is made. Under flexibility come the ideas of using the tooth brush for making candle stand, for inserting cotton rope in piping of 'pyjama,' as handle for dissection needles, as a stirrer, as a spatula and so on.
3. **Originality**	**Originality** It is the ability to come up with uncommon but appropriate responses. Such unusual uses of tooth brush are: – Spraying paint by running it over a wire gauze. – Making coat hanger hooks. – U clips, hands of a clock, and so on.
4. **Elaboration:**	**Elaboration:** Coming to the ability to elaborate, tooth brushes can be shaped and joined together by nuts and bolts to form skeletons of dolls. Heated wire pieces can then be pierced through them to form ribs, fingers, tails, etc. Cinematic and dramatic version of legends are the best examples of elaboration ability, i.e., ability to implement ideas – put them to use.
C.B. (Under ' Elaboration') – Problem Sensitivity – Sense of humour – Serendipity.	Tree more traits of considerable importance are: – Problem sensitivity – Sense of humour and – Serendipity…
4-a: **Problem Sensitivity**	**Problem sensitivity** is ability to sense the problem, to understand what the real problem is, i.e. what are its causes and its effects.

Procedure	Statement
Exercise No. 31. Ask participants to give 2 or 3 such ideas, if any.	Propose 2 or 3 ingenious inventions yet to be invented but which are needed now. The author suggests the following proposals: – Wind force to drive bicycle. – Water repellent wind shield for automobiles. – Swiveling head light(s) for automobile to illuminate curved roads. In thinking of improving function of a tooth bush, a creative person may use new material for bristles that can be sterilized. Or He may provide hollow handle for storing tooth paste and retractable brush head to protect the bristles from dust and fumes. Sensitivity implies not only originality but also evaluative ability. It enables a person to become aware of causes, effects and percussions before putting the idea into use.
4-b: **Sense of Humour:**	**Sense of humour:** Sense of humour is also a significant trait associated with creative intelligence. Humour is appropriate in congruity which carries the opposing elements found in most of the definitions of creative behaviour. It is something like marrying playfulness with seriousness, fantasy with reality, rational with irrational, etc. Best examples of such creative behavior is found in the cartoons (of R.K. Laxman, Shankar, etc.)
4-c: **'Serendipity'**: Story.	**'Serendipity':** Serendipity **is attributed to Horasses Walpele and his story of Princesses of 'Serendipity.** **(Contd.)**

Procedure	Statement
	These princesses had the happy faculty of making unexpected discoveries of things they were not looking for while looking for something else. **Serendipity depends on the awareness of relevancy in accidental happening.** Instead of writing off accidents or happenings as irrelevant, a new meaning related to the goal is searched in it. The famous examples are discoveries of: – X-rays. – Laws of gravity – Antibiotics, etc.
Participants to narrate one such example, real or imaginary, to indicate this trait. Clean C.B	This made Luis Pasteur coin the maxim: **"Chance only presents itself to an alert, prepared mind."**
Problems for Practice Brief the participants about Problems for Practice	Now, a word or two about 'Problems' to be selected for our practice. You might have already thought of 1 or 2 problems or your Seniors might have selected the problems for our practice in this programme. Please ensure that the problem should be describable in about 5 minutes (Max.10 minutes). It may be one of the following type (related to a product or process): – A small assembly. – A measuring device/Instrument. – Job requiring some new process.

Procedure	Statement
	Whatever be the problem you select, it should be from your area of working or one about which you have reasonably sound knowledge.
	Please bring with you drawings, sample, model, Specifications and small hand tools, whatever is required for demonstration.
	Hand tool may include: Screw driver, cutting pliers, spanner(s), tester, etc.
	No riddles or puzzles, please.
	<u>Problem should belong to 'Creative' type, which yields many solutions</u>
	We will work on your problems in 2 teams.
Team work	Each team should discuss the selected problems, and after 'filtering' them, handover the finally selected 2 or 3 problems to me.
	I will collect the problems tomorrow.
	Your problems for brainstorming will be taken up in Session-VI after going through a synopsis of advantages of working in groups or teams.
<u>Subtle warning</u> Talk in somewhat sterner tone.	
<u>Objective</u>: To make the participants to work out the exercises under mild apprehension and tension.	Your management has spent a tidy sum enabling you to attend this course.
Modify the *statement to suit the composition of the group.	I am obliged to give a measured feed back to your management about your enthusiasm in the subject of this session.
CLOSE SESSION.	I would like to base my assessment of your interest on the number of Creative Problems you would provide for practice, and subsequently, quality of solutions

BLAST ----- CREATE ----- REFINE

Blast – To isolate Function.
Create – Alternatives Of Accomplishing Function.
Refine – One Of These Into A Successful Innovation.

7

TRAINING TECHNIQUES OF CREATIVE PROBLEM SOLVING

SESSION – V

(INHIBITORS OF CREATIVE PROCESS)

MENTAL BLOCKS

– EMOTIONAL BLOCKS

– PERCEPTUAL BLOCKS

– CULTURAL BLOCKS

SESSION – V
MENTAL BLOCKS

(Inhibitors of Creative Process)

Procedure	Statement
GENERAL INHIBITORS:	In the previous session we discussed various mental and creative abilities and traits of a creative person.
	In this session we will discuss the factors that hinder free flow of ideas.
C.B. **BLOCKS:**	These factors are called MENTAL BLOCKS or inhibitors of innovative process
	You may agree that a very awareness of causes of any problem leads us more than half way towards the satisfactory solution of that problem.
Exercises to bring out that: – We are not aware of our own Capabilities. – Creativity is dormant.	Our present task is to find the causes that hinder free flow of ideas, i.e. to recognize the attitudes that cramp creativity.
Exercise No. 32.	Please write your name and address by left hand (for right handed person.
Allow 2 minutes.	Please show it to your neighbor.
	Are you able to read and understand?
	OK. Then we go to next exercise.
Exercise No. 33.	How many of you can give more than 5 synonyms of the word 'VITAL' in 2 minutes?
C.B. 'VITAL' Allow 5 minutes.	Please raise your hands

Procedure	Statement
	You may now start writing down as many synonyms of 'VITAL' as you can in the next 5 minutes.
	Is there anybody who got less than 5 synonyms?
Majority will score more than 5.	
Additional Exercises are given in Annexure-13/V.	
	What inferences can be drawn from these 2 exercises?
Get a few views.	When our right hand is injured, we eat food by left hand.
Sum up.	Eating is so essential for survival that ordinarily lazy left hand suddenly becomes active and takes place of injured hand.
	But why not in normal times?
	We seldom realize that the same left hand can also be trained to perform various functions which the right hand does in normal times.
	What is said about left hand is true of our other physical and mental capabilities.
	We are not just aware of our own potentials.
	(When I asked you to give more than 5 synonyms for 'vital', most of you were not sure in the beginning.
	But see the result when once you stated writing down.)
	When we face a new situation or problem, we just give up without attempting to solve it.
	This only shows that we don't have confidence in our own capabilities.
	We undermine our creative potentials by self discouragement.

Procedure	Statement
	You might have seen children running on 4 ft. high walls. They are confident of their agility to jump down safely in case they lose balance. But we the adults curb their enthusiasm in acquiring such skills.
	Result? – In due course, they lose confidence of even standing still on the same wall.
	We have seen earlier how education, training and circumstances force us to use judicial faculty so frequently, how it paralyses curiosity and almost habitually we react to new idea with a big 'NO.'
	This situation is reflected in the comments like:
C.B. (Under 'General Inhibitors) 1. Ignorance of capabilities. 2. Self-discouragement.	– I know, it won't work. – Why change when it is working OK? – What is wrong with this? – Top boss won't agree. – Are you crazy? And so on.
	In the words of Osborn "Judicial thinking is largely negative."
	Premature judgment douses creative flames and even washes away the ideas already generated.
C.B. (Under General Inhibitors) 3. Judicial/critical attitude. Distribute Appendix: AP-O3/V. (67 ways to stop creativity) Allow few minutes to go through. No comments at this stage.	Here is a list of off-repeated responses to new ideas or change proposals.

Procedure	Statement
MENTAL BLOCKS	
	In addition to the three General Disabilities recorded on the board, we encounter many more mental hurdles on the path of creative voyage.
	Scientists have classified them into 3 broad categories, namely:
	– Emotional Blocks.
	– Perceptual Blocks.
C.B.	– Cultural Blocks.
MENTAL BLOCKS:	
I. Emotional Blocks.	
II. Perceptual Blocks	
III. Cultural Blocks.	
I. EMOTIONAL BLOCKS:	Let us now understand what they are.
	First, Emotional blocks.
	Galileo, the inventor of Telescope and a great scientist suffered mental and physical torture for 9 long years before his death in prison.
	The orthodox astronomers refused to look through his telescope and ultimately succeeded in not only banning his telescope but also making him recant his discoveries.
Give deliberate, long pauses between questions so as to make implications clear.	
	How do you explain the behavior of the astronomers?
	What made them hate the original thinker?
	What prompted them to torture the great experimentalist?

Procedure	Statement
Bring out: – prejudice – Jealousy – Closed mind. – Vanity. – Rigid views.	
	If you glance through the list of excuses people advance in preventing change, you may be able to recognize several responses coming from a 'closed mind,' 'jealousy, and vanity.
	Well, that is what the human nature is!
	Most of the great ideas were laughed at when first suggested.
	The great creative minds were jeered, humiliated and even tortured to death.
C.B. (Under 'Emotional Blocks') 1. Prejudice 2. Jealousy 3. Conceit.	Such reactions stem from what is called 'Emotional Blocks.'
	These are external factors and we must recognize them and guard against them.
	But be warned! Greater enemy lies within.
	We must ask searching questions to ourselves to become aware of our own negative attitude.
	– Am I receptive to other's ideas? – Do I encourage others to suggest better methods? – Do I seek or hate others advice? – How often do I use the words 'buts' and "Ifs'?

Procedure	Statement
Case Study-1. (Shot blasting equipment)	Manufacturing process of a newly developed 'split pipe clamps' (G.I. castings) prescribed shot blasting before painting. The main chamber of existing equipment requires urgent replacement. Contractors quoted delivery time of 2 months. If no alternative is made within 8 or 10 days, the delivery commitment will be thrown out of gears. To overcome this problem, engineers from design, methods and manufacturing formed a task force among themselves, and a determined bid to meet the challenge, decided to manufacture the worn out parts in the shop, with one or two added features. The design was evolved in a day or two, and in-house fabrication would have taken 3 or 4 days, leaving enough time to catch up with the time.
Exhibit drawing of the equipment, preferably a transparency, vide Appendix: AP-04/*V*. For transparency, Over- head projector is required. Clear doubts. Say everything was OK, but don't point to the filament lamp.	
	How do you rate the initiative and enthusiasm of the team?
Discuss briefly.	What is the unusual or striking feature of the design?
	Did you notice any abnormal feature in the design?
	Providing filament bulb in shot blasting zone bits all logic!
Bring out: Electrical Lamp.	What could be the reason behind inclusion of such a ridiculous feature?

Procedure	Statement
	Did you notice any abnormal feature in the design?
	Providing filament bulb in shot blasting zone bits all logic!
	What could be the reason behind inclusion of such a ridiculous feature?
Bring out: – Anxiety to meet the dead line. – Tension. – Confusion. – Over enthusiasm to please Higher-ups. **C.B.** (Under 'Emotional Blocks) 4. Anxiety. 5. Over-motivation. **Exercise No. 34**.	The emotional blocks that came in way of rational thinking were 'anxiety' and 'over- motivation.' One more example of this type. A Multi-National Co. manufacturing washing powders had a roaring business word wide. When they formulated a new and powerful detergent, they put up large hoardings in 3 frames all over the world
	1st frame (from left) shows a woman holding a dirty cloth, with a tub full of new detergent foam in front of her. 2nd frame shows the woman dipping the cloth in the tub. 3rd frame shows her holding clean and bright cloth.
Pause for a few minutes and announce.	Within a few days, their sales dropped drastically in the Mddle East, Afghanistan and even in Pakistan. Why?
Point to Anxiety, Over Motivation on C.B.	Arabic, Persian and Urdu scripts are read from 'right' to 'left.' Reason: Anxiety. To be the first......, Over motivation.

Procedure	Statement
Exercise No. 35-(a). (Home Work)	Divide a square in 4 parts of equal areas.
Trainer to note: **Shapes of parts is not mentioned.**	*I will collect your answer sheets tomorrow.*
(Solutions in Annexure-16-A/V)	
Exercise No. 35-b.	
– Draw 2 concentric squares on C.B.	Divide 2 concentric squares in 4 parts of the same shape and equal areas.
– Allow 5 minutes. Collect the sheets and keep them without seeing them.	
– Draw the following figures on CB to indicate the common solutions.	
	If your solutions are other than these three, please raise your hands.
	You may be wondering as to what is so great about this problem and what is wrong with solutions.
Pause for response.	By the by, can you tell me: how many solutions are there to this problem.
	Well, all that you have to do is to count number of 'POINTS' on any line and join them with corresponding points on the opposite lines.
Draw this figure on CB	
	By this time, I hope you are convinced that there are infinite solutions to this problem.
	If you stretch your imagination further and ask yourself:
Pause for expected responses like:	– Why to join the corresponding nearest points?
– Not practical.	– Why not point A to b, B to c. and so on?
– When easy methods are available, why complicated shapes.	– What if the points A-a, B-b, C-c & D-d are joined by curved lines or 'zlg- zag' lines?

Procedure	Statement
	Reasons that might have stopped you from suggesting new and unusual shapes were:
	– Your liking to tread the beaten path.
	– Your allergy to ambiguity/uncertainty.
	– Your fears:
	– Fear of looking foolish.
	– Fear of ridicule.
	– Fear of failure, which made you avoid risks.
	– Fear of being different in suggesting the so-called complicated solutions when simple and easily implementable shapes are available.
	Often we hold back even the most promising ideas due to fear of failure and punishment.
C.B. (Under 'Emotional Blocks') 6. Fear. 7. Allergy to ambiguity. Also point to 'Anxiety.'	There was one more factor that limited your solutions, namely, 'Anxiety' – to impress that you were first in solving the problem.
Case Study-2 (Convenient clamps) Distribute Appendix: AP-05/V Allow time to go through the proposal and reaction to suggestion. Clear doubts.	
	Why this group felt hurt? What type of attitude is this?

Procedure	Statement
C.B. Under 'Emotional Blocks': 8. Negativism. 9. Intolerance.	Negativism is often a product of conceit. When a person can't follow or tolerate other's suggestion, it reflect in such reaction.
C.B. **II. PERCEPTUAL BLOCKS:** – Perception: **Exercise No.36.** **C.B.** Get 4 or 5 different interpretations. **Exercise No. 37.** Stop them writing after 3 minutes	Another set of mental blocks of which we are seldom aware, are called 'Perceptual Blocks Let us first try to understand the meaning of 'perception.' What is your interpretation of this figure? Imagine that you have written a best -seller treatise and the publisher want you to furnish one sheet 'Bios and Burps.' How do you go about it? I am sure you would like to first eliminate all the failures and recollect all the successes spread over 2 or 3 decades of working life and condense them into a dozen or so sentences. You may now commence writing your bio-data.

Procedure	Statement
Let 2 or 3 participants express their feeling briefly. Make a mental note of differences in concept of bio-date.	What was your feeling before commencement of write-up?
Exercise No. 38. (Story of 2 carpenters)	Two job-less, poverty sickened carpenters went to a city in search of work. After wandering through lanes and by-lanes of industrial areas for weeks without any sign of luck, they finally got stuck in their track on seeing a 'wanted' notice on the gate of a big furniture factory which proclaimed: WANTED URGENTLY 3 CARPENTERS. Now, imagine yourselves to be those impoverished carpenters and tell me what would have been your reaction to that notice.
Discuss for a few minutes. Then disclose carpenters' response.	The first carpenter bemoaned: What a pity, we are only two. The second carpenter said: I will go and fetch Narayana from our village. Meanwhile, you hold on the opportunity.
Pause. <u>C.B.</u> Point to 'Perception.' On CB **Pre-conceived views/beliefs.**	Well, such is the magic of perception. Different people attaching different meaning and interpreting the same situation in different ways.
Exercise No.39	Assume that your friend is blessed with a daughter and he wants you to suggest at least 5 names for the baby starting with 'L'

Procedure	Statement
	Please write down and read out when asked for.
After 5 minutes, ask participants to read out the proposed names. Invariably the names will be suitable for a Hindu girl. * Depends on response.	
	* Why not Lina, Luicy, Lilly, Latifa, Laila?
	The names suggested by you are based on your preconceived notion that your friend is a Hindu.
	The story of Dr. Modi reveals the same perception.
	Bit by bit, this trait becomes a habit and leads us to rigidity of stereo–typing.
	– Only males can be efficient surgeons. – Missionary's ultimate aim is conversion.
Exercise No. 40. Rapid- fire questions.	– All traders are tricky, and so on.
	Please answer the following 5 questions spontaneously: – What is the colour of leaf? – How many alphabets in your mane? – How do you ensure that an article is made of genuine stainless steel? – What do you do after sighting a snake in your house? – How many heads to Lord Dattatreya?
Make mental note of the answers and comment as appropriate	
	Some people believe that God has 1 head and 4 hands.
	Many people believe that God has no form, and some reject the very concept of God.

Procedure	Statement
	The members of 'Flat Earth Society' believe that the Earth is flat.
	Disciples of some God-men believe that "Rolex" watches and gold chains can be produced from thin air—no material, no machinery and no technology.
	Some have strong faith in hell and heaven, Rahukalam, black cats and so on.
	These are relatively harmless beliefs or dogmas.
	On the other hand, there are well educated people with strong preconceived ideas.
	– All snakes are poisonous and must be killed, although 80% are harmless.
	– Stainless steel is non-magnetic, although most of industrial stainless steels are magnetic.
	The glass is too brittle to be a structural material – but now very much in demand as FRP.
	The Earth is the centre of our System. This belief held up much progress in science.
	Such beliefs are harmful and misleading.
	Inability to overcome pre-conceived view-points is a major perceptual block to creativity.

C.B.
 	
II. **'Perceptual Blocks'**
 	
1. Beliefs(Rigid Views)
 	
  – Harmless
 	
  – Harmful	

Procedure	Statement
Functional Fixation.	
Exercise No. 41.	What are the functions of a filament lamp?
Answers are likely to confine to a single function: 'Remove darkness.'	
	Your ideas about the function of electric lamp are based on 1 or 2 of its traditional and primary uses.
	You know electric lamps are used in photo studio for drying prints, and in hatcheries for incubation.
	Infra-red lamps are used for curing epoxy based insulation of large electrical machines.
	They also find use in Physio-therapy.
	Magicians gobble down goblets of electric bulbs to entertain people and earn their keep.
	No celebration is complete without garland of decorative bulbs.
Laudable Story.	An enterprising team of chemistry teachers of Kendriya Vidyalaya, New Delhi, has developed a number of apparatus for their chemistry laboratory, using fused bulbs of electric Lamps.
An alternative exercise is:	
Some unusual uses of empty glass bottle of syrup:	
– As a mould for circular holes in wall.	
– For shaping gully traps.	
– Boys playing 'Bowling' game.	
– As a missile, for self defense.	
– Also, artistic work like ship or flowers In bottle.	

Procedure	Statement
	These and our earlier examples on uses of rolling pin/brick bring out ability to visualize multiple uses of a given object.
	But rarely do we think beyond primary functions:
	– Chalk for only writing.
	– Coal for only burning.
	– Milling machine only for milling, and so on.
C.B.	This attitude is called '**Functional Fixation.**'
2. Functional Fixation	
--	---
Inability to see problems from their true perspective.	
Case Study-3.	A Production Manager with more than 12 years of experience in Engineering Workshop had bought a plastic tub for giving bath to his baby. Since the bathroom was small, he had thought of hanging the tub on the wall after use. He knew that I had a small hobby workshop in my house. One evening he and his wife came to our flat with the tub and mentioned the purpose of their visit, leaving the tub near my work table. After usual formality of serving tea, my friend politely reminded the purpose of his visit.
(Plastic Tub)	
	'It is already done,' said my wife.
	'Oh! You also know how to operate drilling machine, surprisingly enquired my friend, adding 'but I did not hear the sound of drill machine.
	Is the problem setting clear?
Pause for response.	What, in your opinion, is the real situation?
	The thing is that when my wife was making tea, she heated a poker and pierced it through the side of the tub near its rim.

Procedure	Statement
(Functional Fixation. Holes are made by drilling only) Get 2 or 3 responses and push on.	What is the message of the incident?
Exercise No. 42. Additional exercises are given in Annexure – 14/V Draw the figures quickly.	Next exercise relates to answering 2 questions in one minute in the form of sketches. As an example, if I ask you to draw a figure of a star or a tumbler, you may draw like this:

Wait for a minute
 	
Then draw a figure of a needle like this:
	Now, the first exercise for you is to draw the figures of:
	   1. Needle.
	   2. Cross section of a pipe.

	I presume all of you have drawn the figure of needle like this:
Point to figure on the board. Draw a machine needle	Do you know that more than 85% of clothes are stitched on sewing machines and 90% of needles manufactured in the world have this shape.
Draw the cross section of a pipe like this: 	

Procedure	Statement
(You should have a piece of square pipe, either real or a card board model.) Show a piece of square pipe like this: 	Well, you know this is also a pipe. There are pipes with rectangular cross section, elliptical cross section, triangular cross section and hexagonal cross section.
Case Study – 4. Read out (Grinder and goggles)	Mr Adinath, aged around 45 and very considerate father of 3 children, is a highly skilled tool grinder working in the precision machining shop. He is a reasonable worker. All through his service, he observed all safety regulations meticulously. But one day he was found to be working briskly on a tool and cutter grinder without putting on the goggles. Noticing this, the shop supervisor first reminded him, and then during his second round, advised him somewhat sternly but to no avail. This continued for 2 days. On third day the supervisor lost his cool and took Adinath to task in front of other workers. Then he reported the matter to the Manager. Supervisor was worried that something happens to the Grinder, production would be adversely affected and he will have to face the music. Manager called Adinath to his cabin and reprimanded him severely. Adinath was upset, felt let down and was heard saying "nobody cared' for him. He almost stopped working.

Procedure	Statement
	The supervisor was now more worried and did not know how to handle the situation.
	Is the problem clear?
Clear doubts, if any.	If you were in the position of that supervisor, how would you had handled the problem?
	What was the real problem?
Get 2 or 3 views.	Finally the matter was reported to Works Manager who invited Mr. Adinath to his office and put him at ease by having tea with him and paying compliments to his high skill, etc., before coming to the point.
	It so transpired that Adinath had a boil behind his ear and the goggle strap was pressing on the apses causing unbearable pain.
	What is the message of this case study to us?
Pause.	These exercises and case studies illustrate how habits and training harden our line of thinking.
Sum up:	– How we try to solve problems habitually by previously successful methods.
C.B.	– How we jump to conclusion – our inability to differentiate between cause and effect.
Under 'Perceptual Blocks':	
3. Inability to see problems from true perceptive;	
– Habits.	
– Training.	
– Confusion between cause and Effect.	
– Jumping to conclusion.	

Procedure	Statement
C.B. Under 'Perceptual Blocks': 3. Inability to see problems from true perceptive; – Habits. – Training. – Confusion between cause and Effect. – Jumping to conclusion.	
III. CULTURAL BLOCKS: Watch: Stony silence prevails. Wait for one or two excuses like: – Fear of getting 'involved' with police. *(Jenny = female donkey) Pause.	You might have seen many an epileptic fallen down unconsciously on footpath and by roadside with lot of spasms and foams oozing out from his mouth. He was convulsing, cringing violently. He was in great agony. What have you done (or would have done) on seeing such a patient on the road? Nothing. You have done nothing to help that unfortunate fellow citizen. Why? It is believed that there is no cure for epilepsy but Ayurveda doctors say the intensity and frequency of attacks can be reduced considerably by giving 3 doses a day of *jenny's urine. How many parents and relatives of an epileptic would care to treat him by this method? None—or one in million? Why?

Procedure	Statement
	You might have seen some victim(s) of 'Hit-and Run' type of road accident(s) needing immediate medical attendance but you have done nothing expect probably taking a photo an collecting information – later on for telling a juicy story as to how you were the first to witnessed the tragedy and so on. You might have also seen municipal water taps leaking heavily. But you did nothing about them. Did you? Why not? Well, that is due to our up-bringing, our training and our culture. Result: complacency, apathy and laxity.
Case study No. 5. **(Caps & Plugs)** *NB = Nominal Bore Distribute: Appendix: AP-06/V (DRG. Of Plugs & Caps)	In a hydraulic system, fluid flows through carbon steel pipes of NB* ½," ¾" and 1," having male and female threads at their both ends. Before dispatching to sites, pipe ends are closed with threaded caps and/or plugs. The end closures are machined from solid aluminum rod of Ø 30, and the average cost of each piece is Rs. 58.20. Cost of annual requirement works out to Rs. 6.28 Lakhs. Let me add: The end closures have one-time use. They are removed and thrown away at sites before re-assembly of the consoles. Can you suggest some alternatives to reduce cost? When I made enquiries about functional requirements and other possible users of such end closures, I came to know that in 3 other shops, similar requirement was being met by Plastic caps and plugs, each costing about Rs.1.75 i.e. about 3% of the cost of Aluminum items.
Get 2 or 3 ideas.	

Procedure	Statement
Bring out: – Compartmental working. – Lack of knowledge. – Laziness/Laxity.	Just imagine where an item worth Rs.3 can serve the purpose, Rs.100/- are being spent to achieve the same function. What could be the possible causes of such a situation? This laxity stem from our starved sensibilities – which have been dulled right from childhood by constant exhortation like: – Don't do this and don't do that. – Don't be emotional, be practical. – Don't poke your nose. – Be precise.----------etc. The spirit of enquiry and exploration has been displaced by over specialization and yearning for stable career. Result: Apathy, lack of enthusiasm to acquire knowledge, etc.

<u>C.B.</u>
Under ' Cultural Block':
1. <u>Complacency, lethargy and apathy</u>.
 	
    (Conformity)
    (Resistance to change).
    (Mental laziness)
    (Don't rock-the-boat attitude.
    (Risk-free life). etc.	

<u>**Exercise No. 43.**</u>	*Please draw a sketch of a wheel –wheel of any vehicle.*

Procedure	Statement
Allow 2 minutes. **C.B.** Draw 1 or 2 figures like this: **Exercise No. 44** Discuss about 5 minutes. **Exercise No. 45.** Get a few opinions. COMMENTS: **C.B.** Under 'Cultural Blocks': 2. Conformity and Dependence	I presume all of you have drawn circular wheels like this: I have a suggestion to make. Please react in normal way. My suggestion is that all Hindu women should stop wearing 'Mangalsutra' and ' Kumkum.' What is your opinion about introducing sex education in Co-education High Schools? Your design of a wheel is manifestation of your excessive dependence on traditional wisdom. Ever since the wheel was invented, nobody had paused to ponder over it till recently as to why a wheel should be circular and why not triangular or square. An American inventor did precisely the same thing and incorporated triangular wheels in all-terrain armor vehicle that takes marshy lands and rocky roads as easily as to plain roads. The idea owner is now a billionaire. Your reaction to my suggestion on 'Mangalsutra' reflects your touchiness – I mean, fear of rejection by your society

Procedure	Statement
	Consciously or sub-consciously, you are concerned too much about your personal security with your group, your clan, your society, and would not dare to question authority of the scriptures and peers.
C.B. Under 'Cultural Blocks':	
3. <u>Social pressures</u> 4. <u>Reliance on authority</u>.	Those social tensions, resistance to change and excessive reliance on fate (–fatalism–) drive one almost to the point of inaction. It is nothing short of a sort of 'cultural shock' to creative efforts. We are the owners of these blocks to varying degrees and we are aware of them. But, surprisingly we do precious little to acknowledge them and to free ourselves from their clutches. In fact, we seem to defend them jealously as something bordering to virtues. In the next session, we will discuss strategies for unlocking potentials and factors that foster divergent thinking.
<u>Distribute handout: CPS-04/V.</u> 'The things they say about…'	
TEA/COFFEE BREAK	

8

TRAINING TECHNIQUES OF CREATIVE PROBLEM SOLVING

SESSION — VI

– PROBLEM FINDING

– FACT FINDING

– IDEA FINDING

SESSION – VI
PROBLEM FINDING, FACT FINDING AND IDEA FINDING

Procedure	Statement
– Collect Answer sheets for Ex. No 35-a	Here are few examples.
Collect problems from participants, study their suitability for creative attack. If not, delete them with permission of the contributor(s). Open Session.	 In the previous session, we discussed various ABILITIES: Mental abilities and creative abilities, and also the 'INHIBITORS' that block our creative hard or flexi-output.
Ex.49 & 50 are directed towards generating awareness for improvement, improvisation, innovations. They should trigger an inner urge and strong will to identity problems for creative attack. (Problem Sensitivity). **Problem Finding: Warm-up Exercises.** --	

Exercise No. 46.
 	
C.B. (Record ideas.)
 	
You may give a lead by stating that you would like:
 	
(1) To have a pad lock without levers & coiled spring but with a key.*
 	
                OR
 	
(2) To invent a hard or flexi-board on which matter can be written by finger or wooden stick (No chalk, No dust, No pen).
	What improvements would you like to make in:
	
	        a) Pad lock (for common use).
	        b) Gas stove, or
	        c) Shirt button? (Any one).
	
	*(The author has one pad lock like this)

Procedure	Statement
Exercise No. 47. **C.B.** Get at least three such "felt-need" problems and record on C, B.	What inventions you would like to have to solve everyday problems. Your contribution should make one thing clear: Knowing what we are looking for helps us to recognize it when you look for it enquiringly. Now, a problem for creative attack can be recognized in two ways: – By redefining the problem consciously, and – First by stating the problem as broadly as you can, and then by breaking down this problem into a number of specific problems.
Exercises on: – Problem redefinition/Restructuring; – Broadening a problem; – Breaking down the problem into sub-problems.	Let us examine the advantages of specifying our problems consciously by using synonyms in multi-expression of the problems, and of spitting a problem into specific sub-problems.
Exercise No. 48.	Your friend has passed a coveted degree exam, standing first in the State. How would you begin a congratulatory letter to him?

Procedure	Statement
<u>Allow a few minutes:</u> Invariably, he response will be something like this: – Hearty congrats … – Immensely pleased to know… – Delighted to know… – It was a pleasant surprise to know… etc. Without comments on the participants' contribution, continue further.	
	Now, I will rephrase the same problem. "In how many ways can you convey your congratulations to your friend?"
<u>Allow 5 or 6 minutes:</u> Expected response would now run something on the following lines: – Meet him personally; – Send a standard printed card; – Talk to him on telephone; – Send a letter; – Convey your message through somebody visiting his place; – SMS or email and so on.	
	The first statement of problem limited the choice of communication to only one method viz. 'writing.' By re-phrasing the same problem, we could get a multiplicity of communication ideas.

Procedure	Statement
Exercise No. 49. (A Story) Allow few minutes: Anticipated suggestions would be as follows: – Serve porridge (kheer) of… – Mix egg or chocolate or 'Viva' or 'Horlicks' to milk – Coax by offering chocolates or new dress every month. – Mix syrup in milk/shake – Hold out threat of punishment like… "No permission to play with friends unless…" "No movie, No TV, No picnic…unless…" etc.	A mother of 7 year old boy has been complaining that her son does not take milk. In fact the dislike the very sight of a milk glass. Could you please help this lady by suggesting as to how to make the boy drink milk? Let me tell you who solved this problem and how. In a "coffee party" when the mother was narrating her woes to her friends, the maid servant, who was fed-up of listening this woe repeatedly, asked suggestively; "Why do you want him (the boy) to drink milk?" Provoked by the unexpected question from the most unexpected quarter, the ladies took it as a challenge to their education, status, etc. and thought aloud. Yes, why should milk be trust down the throat of the boy? Why indeed?" They <u>re-defined</u> the problem by answering the maid's question like this: "Milk is required to provide body building proteins and nutrients."

Procedure	Statement
	Correctly stated, the problem then became – what else can supply these nutrients?
	Well, there was a long list of useful suggestions for the mother and the child to be happy thereafter
	The above two examples should make it clear that problem is not "how to make a child drink milk" or "how to write a congratulatory letter" but how to specify our problems consciously.
PAUSE:	A problem well stated is half solved.
	Coming to the 2nd aspects, namely, splitting of all inclusive problems into specific sub-problems, let us take one or two examples.
Exercise No. 50.	A manufacturing company is plagued with low productivity.
	A broad statement of the problem as to "how to improve overall productivity" in the company will lead us no-where.
	Instead, if we break it up into sub-problems, it can help us in focusing our attention on specific areas.
	Thus, we may ask ourselves:
	– In how many ways can we reduce inventory of:
	a) Raw-materials;
	b) Consumables;
	c) Finished goods?
	– In what way can we increase utilization of:
	a) machinery & plant;
	b) materials;
	c) labour?
	– In how many ways can we cut expenses on Energy?
	– In what way can we increase turnover or Attract more customers? And so on.

Procedure	Statement
Point out to "ideas" recorded in response to Ex.49 & 50.	In fact, each of the sub-problems can be further sub-divided into several specific questions for creative attack.
	Let us take one more example of problem stated in broad term:
Exercise No. 51.	How to improve slums in cities?
	What are sub-problems of this broadly stated problem?
Expected sub-problems: – Sanitation. – Drinking water. – Roads & Trees – Schools & Play grounds – Toilets and Biogas plants, etc. No discussion, no comments and no elaboration.	
Lecture method.	Please remember that every problem for Creative Attack is a question but every question is not necessarily a problem for creative attack.
	Do we need to increase or reduce staff? Such questions may invoke a mere: "Yes" / "No" / "Perhaps" answer. Such questions are a matter of judgment and not for a creative attack.
C.B.	
1. PROBLEM FINDING:	Where do you find problems? Wherever man exists.
Briefly review types of problems.	Common man and most of Scientists and Technocrats heavily rely on judgmental or logical system of solving problems.
	We have also seen that Analytical problems yield only one correct solution.

Procedure	Statement
	Creative type of problems has a large number of solutions, none of them may be absolutely correct but most of them are workable, now or later.
	With this background, there should be no difficulty in Finding Problems For Creative Attack.
Copy the questions on a card and pose them in your natural style.	Pose the following questions to yourself. – Can it be handled in an imaginative, innovative way? – Will it yield result in multicity of useful alternatives? – Does it involve a new, unusual way of solving it? – Can it be re-cast or re-phrased to yield novel solutions? If your answer is in affirmative, then you have found a creative type of problem…
Exercises:	Let us carry out some exercises to find out whether they are of creative type.
Exercise No.52.	How to prevent accidents on un-manned level crossing of railway track?
Make a mental note of ideas.	
Exercise No.53. Pause for the first flush of ideas to cease. Distribute Appendix: AP-07/VI. Allow few minutes to scan the list.	List all possible uses of **sand**.
Exercise No.54. Write the digits on C.B. with gaps for signs as in statement column.	Can you make this sum correct? 1 2 3 4 5 6 7 8 9 = 100. You may use any of the signs (+), (–), (×) before the digits.
Just make a mental note.	What is the type of this problem? Is it Judgmental, Analytical or Creative type?
Distribute Appendix: AP-08/VI.	

Procedure	Statement
Exercise No.55. **Go round the table and check whether any other solution was found.** **Draw the solution on CB.** **Exercise No.56.** (Hint for Trainer: Imagine mirror image of AEB standing on AB). **C.B.** **2. FACT FINDING.** Distribute Appendix: AP- 09/VI. ----- (Document Flow Chart and Info. collection)	In trapezium ABCD, CD = 2 AB and angle ACD = BDC = 45°. Divide ABCD in 3 equal parts (i.e. each of the same shape and same area). This type of problems yield only one correct solution. What is this type of problem called? Is it suitable for CPS process? Can you form 5 triangles with 9 match sticks? Please take it as homework. Having practiced the skill of identifying problems for creative exploration for solutions, let us now turn to the next steps in CPS. Second step in this process is to collect relevant information from **authentic** sources. This is called 'Fact Finding or Data Collection Phase.' As an example, in case of a Manufacturing Industry, you will need data from various functional areas as shown in this Annexure-AP-09/VI.

Procedure	Statement
Talk while going round the table.	This list is neither extensive, nor exhaustive. Requirements of information will vary from project to project and has to be obtained in a 'Tailor-made' fashion.
	However, it is worthwhile to repeat the caution.
	"Seek information from the most **authentic** source only" – and not from the so-called knowledgeable persons because the information given by such persons may be based on insufficient knowledge, heresy, guess work, biased mind or preconceived notions.
	Such information will not only mislead and confuse you but is likely to land you in trouble in the future phases of the CPS programmes.
	Can someone pass on his own experience of this sort? (That is, information given by a so-called knowledgeable person has put you in embarrassment).
Get 2 or 3 stories based on participants' personal experiences, in a few minutes.	Secondly, don't saturate data with facts and more facts, but line-up only a few salient facts related to the problem on hand.
Perils of presumptions. (Inadequacy of 'information.' 'Haste' in offering Solution).	Also, be warned of perils of presumptions, and faulty phrasing of problems.
	Please recall the names suggested by you for the baby of your friend, and also your opinions on Adinath.
Exercise No. 57.	In the middle of night, a mother of a sick child found that his temperature has shot up to 105° *and* he is vomiting.
	What ideas would you put forth to help her in this situation? (Let me tell you that her husband is on a foreign tour).

Procedure	Statement
The probable suggestion would be: – Call a taxi or ambulance and take the child to hospital. – Carry the child on shoulder to hospital. – Call the doctor to her residence, etc. (After listening to the suggestions and depending on their nature, your comments should point to **inadequacy of 'information' and 'haste' in offering solutions** to the problem based on previous experience.)	
You may now announce:	The mother is a famous physician, has a car of her own, knows driving and runs a nursing home in her residence and the child is 14 year old, well built lad. The problem is not how to take the child to hospital but how to obtain medical aid.
C.B. **3. IDEA FINDING TECHNIQUES:** – Creativity Development. – Creativity Techniques Lecture Method.	**Development Of Creativity:** The starting point of any improvement, innovation, or invention is generation of worthwhile ideas. But generation of an idea is in itself a complex psychological process and there is no generally accepted theory or the model available to suggest the possible way of idea generation. **– Creativity Techniques:** However, you are advised to be reasonably familiar with various Models & Techniques of Creative Development listed below with brief description.

Procedure	Statement
	(This being an introductory programme on CPS, I will elaborate only the most popular Technique called Brainstorming, after a while.)
Models **C.B.** – Perception Model. – Fusion Model	a) Perception Model involving both sensory and extra sensory perceptions. b) Fusion model in which information is the key concept and is based on the assumption that if a technical opportunity is recognized, innovative ideas spring up to fulfill the perceived need
Techniques: 1. Morphological. 2. Brainstorming. 3. Forced Relations 4. Synectics*.	**MORPHOLOGICAL TECHNIQUE:** It is an analytical tool of ensuring that all possible solutions to a problem are taken into account through a systematic break-down of problem into sub-problems which can then be treated independently. **2. BRAINSTORMING:** It is method of creative thinking based on free association and deferred judgment. (It is described in details after a while.) **3. FORCED RELATION TECHNIQUE:** In this technique forced relationships are established between two or more ideas or objects or elements which are seemingly unrelated according to habitual thinking patterns. Most commonly used approaches are: i. Catalogue technique; ii. Listing technique; iii. Focus –object technique. The elements chosen are considered in all possible combinations as a basis for free association from which novel and practical ideas will emerge.

Procedure	Statement
*SYNECTICS means putting together of diverse elements. It is related to behavioural science.	**4. SYNECTICS *:** It is a method of problem solving where one attempts to simulate the thinking process when the individuals are most creative. The participants include a client who has an unsolved problem and a group of member trained in the method. The procedure involves statement of problem as given by the client, as by the group, stimulation of divergent thinking by putting evocative question which leads to personal analogy, direct analogy, or symbolic analogy in obtaining a solution. **5. WORK SIMPLIFICATION:** Work simplification is a special application of a Method Study where the <u>employees</u> are trained to analyze and improve the work they perform, in a systematic manner. **6. QUALITY CIRCLE:** Its essence is team approach at grass root level where the working level people are encouraged to identify quality problems, suggest improvements and implement the ideas. Quality being synonymous with Value, Quality Circles aim at improving Value of the operation, product or service – in fact, the ultimate aim is to improve 'quality of life' through numerous suggestion. Before taking up Group Brainstorming, let me give you a synopsis of advantages of group decision.
Lecture method. <u>TEAM WORK</u>	Research in the area of group-decision making indicates that such decisions are superior to both the majority of individual judgments contributing to it and average of individual decisions. This is known as 'Consensus Technique.' No decision becomes final which cannot meet with approval of each and every member of the group. For this reason, consensus is very difficult to achieve.

Procedure	Statement
	However, this difficulty can be minimized considerably by treating the suggestions impersonally, i.e. from the team rather than the contributor' view point.
	Please remember that probability of obtaining a good sense level is improved by using more people in taking a decision.
	Keeping these facts in mind, hereafter we will work in teams.
	You will form 2 teams and each team will elect its leader by consensus.
	Please tell me as to who will be the leaders of Team A and Team B.
Make a mental note of the teams and Team leaders.	
GROUP BRAINSTORMING	
	Group Brainstorming is one of the simplest and most effective techniques for generation of profitable ideas.
	It is based on free association and deferred judgment.
	The approach involves statement of problem, first by the Team leader or facilitator, then by the group, and lastly, selecting the best statement. If the client (the problem owner) is present in the session, then he will be the first to state the problem.
	The emphasis is on generation of a good number of ideas in an atmosphere free from constraints, negative responses, critical judgments, and boundary conditions imposed by traditional thought process.

Procedure	Statement
C.B. **Rules of Brainstorming:** 1. Defer Judgment. 2. Welcome Free Wheeling. 3. Aim for Quantity. 4. Combine and Improve Ideas. Read the Rules slowly. Point to Rules on board. Allow brief debate on interpretation of each rule.	 Basic rules of brainstorming are: – Suspended judgment, **i.e. no criticism, no evaluation of any idea.** – Wilder and original ideas are welcome. – Quantity is encouraged. – Cross fertilization freely allowed. Let us adhere to these principles strictly. Let the ideas flow freely, without any hindrance, and judgment on their utility or futility. Please remember the time for evaluation comes later. Secondly, wilder and even seemingly impractical ideas are welcome. Let us also remember that quantity helps bread quality. The greater the number of ideas flowing, the more chance of striking original ideas. Combination of 2 or more ideas yields a new crop of ideas. Very soon, we will proceed to put these principles in practice on problems selected by you. Before taking up your problems for brainstorming, I would request you to read the rules, one by one, and elaborate them.
Control the debate tactfully.	As we have discussed earlier, the first step in CPS is 'Selection of Problem' for creative attack.

Procedure	Statement
<u>Pre-session preparation:</u>	You may find it difficult to agree on the problem to be solved.
1. 'Selection of Problem.'	In such cases, first, Teams should arrive at an agreement and then, prioritize the selected problems for attack.
2. Phrasing of Problem.	Next, the whole group should select 2 or 3 problems from each team that are likely to yield good number of useful ideas.
	Once again, I would like to remind that your emphasis should be on wording of your problem.
	– 'How many ways can we convey best wishes......?'
	– 'How to improve productivity in our machine shop?'
	– 'How to combat downfall in sales?'
	– 'How can we...?' How might we......? And so on.
	Phrasing and re-phrasing of a problem like this will yield multicity of solutions.
	Please also remember that <u>broad statement of problem</u> will lead us nowhere, and how division of broad problem into sub-problems gives good results
LUNCH BREAK	

9

TRAINING TECHNIQUES OF CREATIVE PROBLEM SOLVING

SESSION — VII

BRAINSTORMING

PARTICIPANTS' PROBLEMS.

SESSION – VII

BRAINSTORMING: PARTICIPANTS' PROBLEMS

Procedure	Statement
Sort the problems team-wise and hand over them to the Team Leaders. (Read NOTES on page 105.)	Here are your problems. – Team A will hand over the list of their problem to team B and team B to team A. – After consulting the problem owners, eliminate riddles, puzzles and problems which do not fall under the category of 'creative' type. – Next, teams will allot 'ranks' to the problems. A problem which is likely to yield maximum solutions is ranked "1," and so on. – Now, the whole group meets and 'filters' the problems to select 3 best problems to brainstorm, lists them according to their ranks and hands over the list to me.
– Read out the selected problems slowly with pauses between them. – Brainstorm the first problem, involving all participants as a single panel. – Start brainstorming first problem. – Sit down and record ideas on a sheet of paper. – Video-record or tape-record the proceedings. At the end of allotted time, read out the listed ideas. TEA BREAK.	– All of you should participate enthusiastically in offering solutions, strictly adhering to the rules. – Please keep the pad and pen ready. – Start. Please remember: Duration is 20 minutes. We will take up the remaining problems in the next session.

NOTES ON BRAINSTORMING.

BRAINSTORMING PARTICIPANTS' PROBLEMS.

1. After collecting participants' problems at the end of Session-IV and before conducting brainstorm session, you as the panel leader should ensure:

   a. that between the sessions, you have studied the problems in consultation with the 'owner' of the problem and understood it thoroughly.

   b. that you have necessary 'information' on hand, analyzed and systematically arranged, particularly in respect of a problem related to 'product' or 'process.'

   c. that the copies of 'information' related to 'product' or 'process' are available for distribution before brainstorm.

2. For brainstorming members problem,, you should develop in advance a list of your own ideas so that if the flow of ideas slows down, you can prime the flow by interpolating ideas of your own.

3. You should be well prepared to give hints in the form of 'spur' questions like:

   – Why is it necessary?

   – What else can achieve this function?

   – Where should it be done?

   – Who should do it?

   – How should it be done?

   – When should it be done?

   You may copy the 'spur' questions on a card and use it in brainstorming session.

4. You should kindle the fire of imagination by 'strokes' such as:

– What about? – What if?

– What else? How else?

– Why not eliminate?

– Why not combine?

– Bigger instead of smaller?

– Smaller instead of bigger?

– Why not reverse?

– Why not rotate?

– Why not rearrange?

– Why not shift location?

– Is there something similar? And so on.

This is an endless list of such questions and you should prepare your own list that activates imagination in right direction.

10

TRAINING TECHNIQUES OF CREATIVE PROBLEM SOLVING

SESSION – VIII

BRAINSTORMING – CONTINUED

BRAINSTORMING (CONTD.)

Procedure	Statement
BRAINSTORMING: Participants' Problems.	In the last session we have subjected your first problem to creative attack and got (many) alternatives as solutions.
	Now, please take up your remaining 2 problems: One problem will be brainstormed by Team A and the second one, by Team B, strictly following the rules.
Call the leader of Team-A to the head of the table. (One more member from his team may also join him to assist in recording ideas).	
Request him to present problem up to "Information Phase" only, and to conduct brainstorm session.	
Also ask him to read out the problem on hand, slowly.	
Prompt him to put appropriate 'spur' questions to the panel to evoke spontaneous responses, suggesting alternative solutions.	
– Sit down and record the ideas on a sheet of paper.	
– Video-record or tape-record the proceedings.	
– Allow about 20 minutes.	
– At the end of allotted time, read out the ideas and get agreement of the participant	
Thank the Team Leader (and his partner) and return them to their seats with a word of encouragement.	
REPEAT FOR THE LAST PROBLEM.	We have now 3 lists of ideas for evaluation in the next session

IMPORTANT NOTE: If none of the brainstormed problem was up to the mark OR if about half an hour time is left before closure of the session, take up the following problem:

This is a real problem faced and solved by the author, using novel methods. Please announce that this exercise is based on technique of 'Individual Creative Efforts,' which include passion for solving problems, ability to take quick decisions in associating different elements from one's store of experiences, attitude and aptitude.

Procedure	Statement
CASE STUDY- 6 **(Non-magnetic Flats)**	**Introduction:** Non-magnetic (Manganese steel) flats are welded radial on the Pressing Plates of rotors of Generators, as stiffeners. SS 316 electrodes are used for welding. **Information:** All dimensions are in millimetres (mm). 1. Cross-sections (C/S) of flats are: 8×50, 10×50, 10×70, 12×50, 15×70, etc. 2. Flats get rusted, and can be bent easily like CS flats. 3. Length of cut pieces varies from 250 to 675. 4. Length of raw material is 3500 ± 200. 5. This material is as costly as stainless steel. 6. Route Card (process sheet) says cutting in 'Machining Centre,' without specifying type of machine & cutter. 7. Route card says 'Do not use 'blue' to mark positions of cuts. <u>**In how many ways can you cut raw material into pieces, say, of length 280 mm?**</u>

<u>Problem:</u>
 	
A) <u>Participants</u> in Idea generation:
 	
  Highly skilled Artisans $= 6$
 	
  Supervisors $= 1$.
 	
  Customer's Engineer $= 1$.
 	
B) <u>Ideas:</u>
 	
Use: 1. Power Hacksaw.
 	
  2. Milling M/C with:
 	
    HSS End Mill, Slitting saw.
 	
  3. Shearing.
 	
  4. Gas cutting.
 	
  5. Wire cutting.
 	
  6. Press break, etc.
 	
Nobody suggested "Abrasive Cut off machine"
	All these ideas were rejected for the following reasons:
	
	<u>Ideas-1 and 2:</u>
	
	Manganese steel is a work-hard material. HSS cutters got blunt without cutting.

Procedure	Statement
	Ideas-3 and 6:.
	C/S of sheared cut pieces resembles C/S of cast iron (coarse), and not acceptable. Accuracy of lengths is beyond given tolerance.
	Idea-4:
	Gas cutting makes the ends red hot and changes properties of the material. Hence not accepted.
	Idea-5:
	Cost of wire cutting was more than the cost of a flat.
Solution:	Problem owner, who worked in Foundries, remembered that manganese steel castings were fettled using Abrasive Cut- off Machine. Why not use this m/c for our job?
	Trials were successful.
	Then, routine activities involved and completed were:
	Selection of grade and size of wheel, negotiation with manufacturers of wheels, changing safety guard, adj. stopper, etc.
	Process of cutting flats was successfully developed.
Next problem:	After cutting 100 pieces, inspection revealed that length of about 60 pieces was 283 ± 0.5, as against 280. i.e. 3.5 ± 0.5 more than required length. (Stopper moved.)
	When attempt was made to cut off extra length by abrasive cut off machine, the wheel was bending and giving taper cut. Also danger of wheel breaking and causing grievous injuries to worker was very high.
Ideas:	**How to remove extra 3.5 mm at least cost?**
1. Mill using T-max cutter with carbide tips.	
2. Grind off extra material.	Both ideas were workable and subjected to 'Evaluation Methods,' and found to be laborious, time consuming and very costly.
	At this point, the author pondered over 'Cause and Effect.'
Solution:	Why was the wheel bending while trying to remove 3.5 mm?

Procedure	Statement
	Because there was no support to wheel at open end.
	(Wheel is 3 mm thick and material to be removed is also 3.)
Disclose the solution after getting 2 or 3 ideas.	
Solution:	End pieces (left-over/scrap) were tack welded to give support to wheel, and excess length was cut off, using the same cut-off machine.
ONE MORE PROBLEM:	**Next operation was to machine one of the 4 corners of the flat to R10.**
	How?
	Route card says: Hold 5 pieces in machine vice and gang mill the corner at 45° up to 6 mm, and then grind to R 10.
	Any alternative, cheaper method?
	When cut-off wheels wear out to about Ø 200, they are scrapped.
Solution: **(Short Radius wheels do not bend)**	Rotate Abrasive Cut-off Machine Head through 45°. Clamp the job in flat-position. Use discarded Cut off wheel and cut off corner. (6 × 6) Then grind the corner to R10.
	(It takes 5–6 seconds. Cost of consumable tool is zero.)
	In fact, there were some more problems in machining of these Flats... They were also solved by using similar novel techniques, resulting in a **massive saving of 60%.**

11

TRAINING TECHNIQUES OF CREATIVE PROBLEM SOLVING

SESSION – IX & X

EVALUATION

IMPORTANT NOTES FOR TRAINER ON EVALUATION

i.   In EVALUATION sessions there are:

   a. Several Criteria to be selected;

   b. One Case Study (Pen Holder);

   c. Examples of Matrix Techniques.

   It is advisable to note down the Criteria, Factors and Prompt Questions on cards, giving page numbers where they occur in this manual, and use them in 'Lecture Methods.'

ii. In Matrix Techniques, there are 13 matrixes in Evaluation Sessions. The best way of presenting them in serial order is to make:

   a. Flip Charts on A2 Size Sheets, giving serial numbers; OR

   b. Transparencies suitable for over-head projector; OR

   c. Any other Electronic device.

iii. If you desire to use Chalk Board for drawing matrixes as suggested in the manual, first, practice on paper as follows:

   a. Take 2 sheets of paper. On one sheet, make 4 columns and fill them up, as per 'Procedure' column

   b. In procedure column there are instructions to 'clean' some parts of some columns and then to draw the matrixes. Use the 2nd sheet to represent the board work after its partial cleaning.

iv. Read the EVALUATION chapter several times and cross check that your plan to use cards, flip charts and board work is okay.

v.   Now, you have a good number of ideas to solve a given problem. Evaluation of ideas or alternatives helps us in making a good decision in selecting one of the ideas for implementation. The decision may not be perfect. It is, however, the best one made on the basis of available data at hand.

SESSIONS – IX & X: EVALUATION

Procedure	Statement
OPEN SESSION:	In the previous session we have seen how to be Deliberately Creative and how to generate multicity of ideas.
	Objective of brainstorming session was to **get ideas, and not to solve problem.**
*Edit.	(Three)* brainstorming sessions yielded many ideas that were read out at the end of last session.
<td colspan="2" align="center">**SELECTION OF CRITERIA**</td>	
	By your judgment and experience you are likely to reject some of the listed ideas summarily, being too wild or ridiculous.
	You may also be tempted to recommend certain ideas for straight-away implementation.
Stress the world "subjective."	Rejection or acceptance of ideas based on judgment and experience will be totally **subjective** and will defeat the very purpose of generating a host of ideas.
Stress the world "Objective"	What is needed at this stage is an unbiased and **objective** treatment in the form of numerical method.
	Please give a few examples of things that are subjective and objective
Subjective: Very late, good, bad, excessive, incorrigible, etc. Objective: 1 Hr. late, 20%, ZERO, 99°C, 35 year old, etc.	
	In this session we will practice some important techniques of judging the relative effectiveness of ideas and arriving at the optimum solution.
CLEAN C.B.	

Procedure	Statement
C.B. (1ˢᵗ column) **EVALUATION:** **Objective:** To practice selection of criteria of evaluation and their ranking order of importance.	The factors, attributes, or criteria to be considered in the evaluation will, of course, depend upon: – the complexity of the problem or the function; and – the degree of refinement sought to be achieved in evaluation. Accordingly, the process of evaluation will vary from a simple technique of listing good and bad points to the complex method of MATRIX ANALYSIS. Whatever be the technique, the first step in evaluation is the selection of criteria that must be considered in comparing the effectiveness of different alternatives.

C.B.
Write **'Criteria'** under EVALUATION.
 	
Bring out and record the FACTORS in 2ⁿᵈ column under EVALUATION
 	
   – Quality;
   – Reliability;
   – Performance;
   – Price;
   – Availability;
   – Durability;
   – Serviceability;
   – Ease of operation;
   – Safety;
   – Appearance, etc.
	What factors do you take into account before buying a consumer durable like a bicycle or a pressure cooker?

Procedure	Statement
Statements like "consult Friends," or brand names should be related to some of the attributes listed above.	
	What additional attributes should be considered in case of an electrically operated appliances like a table fan, or a washing machine?
Record additional factors only on the C.B. (like power consumption, capacity, electrical reliability, etc.)	
	And selection of a fridge?
Just listen and push on.	
	So, it is advisable that our first task in evaluation is to prepare a check list of common yardstick with which to measure the comparative merits of ideas.
Distribute Appendix:AP-10/VIII (Pen Holder).	
Case Study-7. **(Pen holder)** Allow few minutes to study the text and illustrations.	Here are some ideas suggested by a group of technician apprentices for a ball-point pen holder.
	What are the possible attributes that are appropriate for evaluating these ideas?
If necessary, use suggestive questions (listed on next page) to bring out about 4 or 5 factors amongst the followings: – Material; – Number of parts; – Number of operations; – Stability; – Safety; – Design cost; – Aesthetic; etc.	

Procedure	Statement
	– How about number of parts?
	– Does it affect the manufacturing cost?
	– Assembly time?
	– Any special tools and skills required to make and/or to assemble?
	– Do you think that string will last long? (durability)
	– What about safety? (sharp edges)
	– Will the tumbler be stable at all times? (stability)
	– Which design needs least amount of material and operations? And so on.
List 5 or 6 characteristics on C.B. (3rd column) and number them as A, B, C, D, …	
	Which of these are very important or less important and in what order?
	Will you please help me in ranking these factors in terms of their relative importance?
Get the ranks quickly and record them against the criteria: A, B, C, D, ….	
The board work now appears somewhat like this:	
Key	

Letter	Factor	Rank
A	No. of parts	1
B	Durability	2
C	Safety	4
D	Convenience	3
E	Material cost	1
F	Design cost	5

Procedure	Statement
MATRIX TECHNIQUES OF EVALUATION	

Procedure	Statement
	Let us examine whether this ranking by 'hunch' or judgment, stands the scrutiny of a numerical method.
C.B. In 4th column, write: MATRIX SYSTEMS: **(Forced Decision Techniques)** Draw a matrix like this: 	
Explain by lecture method.	
	In the matrix systems of determining the 'weightages' is based on the fact that no two factors have exactly the same merit or value i.e. one of the two factors must be more important than the other, of course, to a varying degrees. The difference in importance may be minor, medium or major, and is denoted respectively by the digits 1, 2 and 3
C.B. (4th column, below matrix) Difference in importance: 1. Indicates minor difference. 2. Indicates medium difference. 3. Indicates major difference.	

Procedure	Statement
	The factors are compared in pairs and degree of difference is indicated by the key letter of the more important factor followed by digit 1, 2 or 3, depending on the degree of difference.
Write faintly A2 in Matrix.	Thus, in comparing A with B, if A is more important than B, and the difference is medium, A2 is entered into the square of matrix which is common to Row A and column B.
	Similarly, if D is less important than F and if the difference is major, F3 is entered into the square which is common to row F and column D, and so on.
Write faintly F3 in Matrix	May I mention that the number of comparison (x) to be made in a set of 'n' factors taking 2 at a time, is given by the formula:
	$$X = \frac{n\,(n-1)}{2}$$
	This formula is used in determining the number of matches (X) to be played between 'n' teams (in hockey or foot-ball league).
	In our case n = 6. $\quad \therefore \quad x = \dfrac{n(n-1)}{2}$
	$$= \frac{6\,(6-1)}{2}$$
	$$= 15$$
C.B.: $$x = \frac{n(n-1)}{2}$$ $$= \frac{6\,(6-1)}{2}$$ $$= 15$$	
	Let us now complete this matrix.
Point to matrix.	

Procedure	Statement
PAUSE: Rub off A2 and F3 from the matrix. Ask the participants to compare 'A' with 'B,' A with C, A with D and so on, and to spell out the degrees of importance. Complete the row A. Next, take B and compare it with C, then with D and so on. Continue till all the factors are compared in pairs. Your completed matrix will look like this:	

Your completed matrix will look like this:

	A	B	C	D	E	F
A	A2	A3	A3	E2	F3	
B		B3	B2	E3	F3	
C			C2	E2	C1	
D				D1	F1	
E					F2	
F					F	

The "Weightage Points" for different factors are then calculated by adding the numbers following key letters like this:

A2 + A3 + A3 = 2 + 3 + 3 = 8

B3 + B2 = 5 and so on.

Explain the method.

First, add the 'numbers' following A, then those following B, etc.

Procedure	Statement
Prepare a table in 4th column of C.B. and enter the score in 2nd column of the following table:	

Weightages				
Key	Points (Score)	%	Fraction of 1	Rank
(1)	(2)	(3)	(4)	(5)
A	8	24	0.24	2
B	5	15	0.15	4
C	3	9	0.09	5
D	1	3	0.03	6
E	7	22	0.22	3
F	9	27	0.27	1
Total:	33	100	1	

Procedure	Statement
Point to above 2 tables on the chalk board.	Let us compare the rankings of factors by the two methods we have followed so far.
Compare the ranks given by 'hunch' and arrived at by numerical method.	
If, by any remote chance, the two rankings are exactly the same, give credit for accuracy of "Guestimates" made in the 1st method; and emphasize that such a coincidence is rare. If not, mention that the numerical method leads to objective evaluation	
	Numerical methods leave no room for ambiguity and lead to objective assessment.
	There are several ways of extending these matrix techniques to evaluation of ideas generated in brainstorming sessions.
	Two of them called GRADING METHODS are discribed below.

Procedure	Statement
GRADING METHODS	

<table>
<tr><td>

<u>C.B</u>. Retain upper half of 1st and 4th column and clean the 2nd and 3rd columns.

In 1st col. Below 'CRITERIA:' write:

1.Value-T Chart-I OR (Grading Method)

<u>OBJECTIVE:</u>

To practice use of various grading (numerical) methods of evaluation.

Explain by lecture method.

</td><td>

</td></tr>
<tr><td>

Poor	Fair	Good	V.G.#	Xlent*
2	4	6	8	10

Draw the following matrix on C.B.

(1st and 2nd columns).

</td><td>

One of the simplest ways of evaluation is to first list out the KEY factors or characteristics that are relevant to the case under study and number them as A, B, C, D, etc.

Next, on a scale of 0–10, designate the grades of each factor as:

<u>Grades</u>	<u>Points</u>
Excellent*	10
Very good#	8
Good	6
Fair	4
Poor	2

Next step is to prepare a matrix like this:

</td></tr>
</table>

Alternative	Design effectiveness					Score	Rank	Decision
	A	B	C	D	E			
A-1								
A-2								
A-3								
A-4								

Procedure	Statement
	Each alternative is now evaluated visa-vis the design criteria, and the points entered into the matrix
	Thus, if alternative A-1 offers excellent quality and reliability, good service ability and safety but poor productivity, the 1st row of the above matrix would read like this:

Alternative	Design effectiveness					Score	Rank	Decision
	A	B	C	D	E			
A-1	10	10	6	6	2	34		
A-2								
A-3								
A-4								

In similar manner, compare the efficiency of 2nd alternative (A2) in respect of the design criteria A, B, C, etc. and record the points in the matrix against A-2. Continue till all alternatives are considered. Complete the matrix; Calculate scores; Rank the alternatives; and Note the decisions.	

The completed matrix reads something like this:

Alternative	Design effectiveness					Score	Rank	Decision
	A	B	C	D	E			
A-1	10	10	6	6	2	34	3	Reject
A-2	8	8	10	10	8	44	2	Hold
A-3	2	6	6	8	8	30	4	Reject
A-4	10	10	8	10	8	46	1	Accept

Exercise on 'Pen Holder' (contd.)	Let us apply this method to the alternatives of the Pen Holder, taking the Fig.1 as standard for comparison and using the earlier listed design characteristics. As before, we will work in teams, and each team will take up 5 alternatives.

Procedure	Statement
Note: Composition on teams and number of alternatives to be taken by each team, depend on No. of participants Allow 15 to 20 minutes. Draw the following 2 tables in 3rd column of C.B. When the teams reassemble, ask the team leaders to give the group decisions. Note the "most desired" alternatives and short-list them on C.B. as follows:	Team-1 will evaluate effectiveness of Alternatives 2 to 6; Team – 2 will Take-up 7 to 11; the 3rd team, 12 to 16 and the 4th team, the rest. You may complete this exercise in about15–20 minutes.

Sl.No	Alternative no.	Decision
1.	A-1	
2.	A-2	
3.	A-3	
4.	A-4	

Procedure	Statement
* Modify to match with No. of teams. All participants as a single group to evaluate the short-listed Alternatives.	We have these (4)* short-listed alternatives for final evaluation, and all of you will form a single panel to complete the exercises

Alternative	Design effectiveness					Score	Rank	Decision
	A	B	C	D	E			
A-1								
A-2								
A-3								
A-4								

Procedure	Statement
	How do you rate alternative A-1 as regards the quality?
	– Excellent?
	– Good?
Enter the points corresponding to the grade in the appropriate box in the matrix.	
Repeat the questions in respect of other criteria as applicable to alternative-A-1, and complete the first row.	
Next, take the other 3 alternatives, one at a time, and subject them to similar enquiry.	
Enter the points corresponding to the grade in the appropriate boxes in the matrix.	
Next, take the other 3 alternatives, one at a time, and subject them to similar enquiry.	
Complete the matrix.	
Ask for decisions, record them in the matrix and read them aloud.	
Clean the lower halves of 3rd & 4th columns of C.B.	
C.B. 1st Col. Under '1, Value-T Chart-I' write:	
2. Value-T-Chart–II (Point Score Method)	In an improved version of grading method which is also known as 'Point-Score' method, the relative weightages of the criteria are first converted into percentages or fraction of '1' like this:
By Lecture Method:	
Refer to the score board in the 3rd column.	
Read out the first 2 or 3 weightages in fraction.	
In 4th column, a new matrix is drawn as shown below	

Procedure								Statement

Alternative	Weightage:	A	B	C	D	E	F	
		0.24	0.15	0.09	0.03	0.21	0.28	
A1		–	–	–	–	–	–	
A2		–	–	–	–	–	–	
A3		–	–	–	–	–	–	
A4		–	–	–	–	–	–	

Procedure	Statement
Record the rated value against A1 (above dotted line) and below A. Continue till all criteria are covered for alternative 'A1.' Repeat the procedure for other alternatives (A2, A3, A4 …).	Next, allocate score varying between 70 & 90 to each criterion, 70 being the least acceptable design effectiveness and 90, the maximum expected. On this short scale of 70–90, how do you rate alternative A1 in respect of criterion A (quality)? What about 'B'?

The matrix now reads something like this:

Alternative	A	B	C	D	E	F	Score	Decision
	0.24	0.15	0.09	0.03	0.21	0.28		
A1	90	90	70	70	80	70		
A2	90	80	80	75	90	75		
A3	70	70	80	90	70	85		
A4	70	80	80	75	70	80		

	Next, the rated values are multiplied by weightage fractions and the scores are written below the dotted line in respective columns Thus, the product of 0.24 and 90, rounded off to the nearest digit is 22 and is written against A1 in the 1st column.

Procedure	Statement
The matrix now reads something like this:	

Alternative	A 0.24	B 0.15	C 0.09	D 0.03	E 0.21	F 0.28	Score	Decision
A1	90	90	70	70	80	70		
A2	90	80	80	75	90	75		
A3	70	70	80	90	70	85		
A4	70	80	80	75	70	80		

Procedure	Statement
	Next, the rated values are multiplied by weightage fractions and the scores are written below the dotted line in respective columns
	Thus, the product of 0.24 and 90, rounded off to the nearest digit is 22 and is written against A1 in the 1st column.
Enter the score 22 in the matrix against A1 below 90.	
Repeat the other criteria and fill up the 1st line (against A1). Add the 'products' and record under 'score.'	The other alternatives are similarly evaluated.
Ask the participants to carry-out the exercise in respect of A2, A3, and A4.	
Get the 'products' and complete the matrix.	
Read out the 'scores' and ask for 'decision.'	
Record decisions in last column.	
	That completes our **Criteria Value-T-Charts Techniques Of Evaluation.**
Final board work is given on next page.	

Procedure								Statement	

Final board work:

Alternative	A 0.24	B 0.15	C 0.09	E 0.21	F 0.28	Score	Decision
A1	90	90	70	80	70	81	2
	22	14	6	17	20		
A2	90	80	80	90	75	88	1
	22	12	7	19	21		
A3	75	70	80	70	85	77	3
	18	10	7	15	24		
A4	70	80	80	70	80	75	4
	17	12	7	15	22		

Compare the decisions arrived at by the two methods.

(They will and should, be the same).

Announce broad outline of programmes for the next session.

Evaluation of ideas generated in Brainstorming sessions

    (Participants' Problems)

We have 3 lists of ideas generated in brainstorming sessions.

For evaluations of these ideas, you will form 3 teams.

Each team will take up one list of alternatives and evaluate their effectiveness

You are advised to:

1. First short-list the alternatives (not exceeding 4 from each list by using simple weighing technique and then,

2. Use criteria value-T-Chart for final evaluation.

<u>Close Session.</u>

12

TRAINING TECHNIQUES OF CREATIVE PROBLEM SOLVING

SESSION – XI

– PLANNING

– DEVELOPMENT

– IMPLEMENTATION

SESSION – XI

PLANNING, DEVELOPMENT AND IMPLEMENTATION

Procedure	Statement
OPEN SESSION. **C.B.**	Let us start our session with the quote: **"If You Fail To Plan, You Are Planning To Fail."** Anonymous. A similar quote coined by Benjamin Franklin reads: **"By Failing To Prepare, You Are Preparing To Fail."** These quotes have a significant relevance to our today's deliberations. We have seen that EVALUATION helps in short-listing of a very few ideas, but not in arriving at the final decision. In Engineering and high Technology fields, even the most promising idea will face the perils of presumptions, if the potential ideas are not developed into acceptable proposal by the stake holders.
C.B. **1.PLANNING & DEVELOPMENT.** – Check lists of criteria Investigation. – Plan of action for: * discussion * overcoming 'road blocks' – Prototype making – Testing/proving – Feed back – Final recommendation <u>Long pause to allow copying of points from C.B</u>	In this session, we will discuss 3 steps of Planning and Development. 1. 1st Step involves preparation of idea-investigation check list of each of the selected criteria, drafting a programme, discussion with specialists, suppliers, customers, and colleagues in other departments, etc. 2. Listing of an anticipated resistance to changes and charting of strategies to overcome the possible obstructions are also a part of this phase. 3.Prototypes, trial testing, collection of feed backs and finally, making recommendation of the most attractive alternative for implementation, are the other activities of this phase.
STEP-1 **1. Preparation Of Check Lists.**	The 1st step in programme planning is preparation of checklists. What should include in the investigation check list for, say, Product Design?

Procedure	Statement
Check List	– What are the required tolerances?
	– Can they be relaxed?
Draw out 3 or 4 of the following:	– Any special tooling or equipment required?
– Tolerances;	– Any corrosive or dusty atmosphere?
– Any special materials & process;	
– Operating environment & life (fatigue, wear)	– Does the function call for duplication of safety provisions?
– Fail-safe features;	– What is the degree of inter-changeability of parts or sub-assemblies?
– Inter-changeability;	– Can any standard parts be used?
– Use of standard components;	
– Optimum mechanical simplicity;	
– minimum weight.	
Bring out few factors like:	What about Quality Checks?
– Customer's requirement,	
– Inspection facilities,	
– Material testing,	
– Pressure testing,	
– Functional Testing, etc.	
STEP-2	
2. Draft a Plan Of Action.	
	With Check lists in hand, we should now ensure that all the questions are answered.

Procedure	Statement
Questions (Copy from a pre-prepared card and record on **C.B.**)	
Who will help? – Designer? – Buyer? – Shop Manager? – Expert?	Who will answer? – Designer? – Buyer? – Supplier? – Any expert or specialist?
Any discussion? – With whom? – When?	Any discussion called for? – with whom & when?
Any additional data? – Who will collect? – When?	What additional information/data will be required? – Who will collect it? – When?
Any prototype/testing? – What cost? – Time?	– Any proving or prototype required? – If so, how long will it take to make it or prove it? – What will that cost? – Are the efforts, time & cost commensurable with the anticipated gains?
STEP-3 **3. Execution Of Plan.**	If the check list is worked through question by question, no difficulty will be experience in preparing a sound and workable plan for further investigation in respect of short-listed alternatives/ideas.
Objective: To secure co-operation in developing ideas to a practical proposition. Lecture Method.	Most important aspect of Execution of Plan is to overcome Road Blocks and win cooperation from the Stake holders for smooth implementation. Please remember, there are people waiting to shoot down your idea bird.

Procedure	Statement
	Your skill in handling human factors will be on severe test, and success or failure of your attempts to bring about the desired change will largely depend on your tact and patient in overcoming the obstruction and securing help of the concerned people.
C.B. (2nd Column.) **ROAD BLOCKS** Ask Participant to refer to Handout AP-03/V. (67 ways ro stop Creativity).	One of the steps in the action plan is "to meet and discuss with designers, buyers, vendors, etc." at mutually agreed time and venue. In the meeting, what type of obstructions do you anticipate and why? You may recall 67 ways to stop creativity.
Bring out 4 or 5 of the Common Mental Blocks used as excuses to prevent change. Allow short discussion to draw out: – Attitude. – Work habits. – Fear. – Laziness. – Ignorance.	Why do people resist change?
	People generally get used to doing things in certain way, in a certain place, at certain time, and with certain people. Now, every change has a smell of un-known danger or challenge to one's authority, expertise, position or status. Sometimes, it is shear laziness.
Excepted response: "Usually resent it."	How do you feel when somebody in your house changes something at home without telling you before hand?

Procedure	Statement
C.B.	We are all inclined to question: Whether it is necessary to change the things to which we are accustomed?
Accept that resistance to change is normal.	
Solution:	One of the best ways of overcoming resistance is to let the people know in advance changes that will affect them. If you could tell them WHY of it, it would be still easier to get them to accept the change.
	Your intention of meeting people, I mean, Specialist, Line Manager, etc. is to "ask for help in developing ideas," and not for imposing change.
C.B. (Copy from pre-prepared card) **GUIDELINES** – Ask for help. – Don't argue. – Listen. – Positive questioning. – Welcome suggestions. – Show respect to experience. – Don't jump to conclusion.	
	These guidelines help you in presenting your case briefly and objectively.
	In a progressive, forward looking organization, approval of innovative proposals does not take much time and efforts.
Final Choice Of Idea/Alternative.	After discussion with the specialist and line managers, comes the time for little introspection, reinvestigation and final evaluation.
	De-bug your proposal of lacunae, if any, and compile a comparative statement on the developed ideas.
Distribute Handout No. CPS-05/XI. (Little Red Hen)	

Procedure	Statement
	Table should indicate as to how each developed idea compares with the existing design and with the other:
	– Estimated Cost.
	– Time and ease of Implementation
	– Effects of selected criteria, etc.
	With comparative data on hand, you may now compile your report for presentation to the representatives of the management in the concluding session of our deliberation.
Invite questions of clarification.	
Clear doubts, if any.	
C.B.	
(Under Planning and Development)	
2. IMPLEMENTATION	Coming to the last phase of this session, you are likely to face tougher task in implementation of your proposals.
	As we have discussed earlier, any change will affect a large number of people or departments to varying degrees.
	For example, if a change in design is called for by your proposals, many things will have to be done:-
	– Preparation of fresh drawings;
	– Replacement of old drawings with new drawings;
	– Ordering of new materials, tools, jigs & fixtures, gauges, etc.
	– Arranging of test facilities;
	– Training of operators, etc.
	Concept of packing or packaging, shipping and marketing may also undergo complete change.
	What problems do you foresee in Implementing your proposals

Procedure	Statement
Invite comments from the group on likely problems of implementation	
Draw out factors relating to:	
– Design, material specifications & procurement;	
– Manufacturing operations (Tooling, J&F, Training);	
– Inspection/Q.C.;	
– Testing (proving);	
– Marketing (packaging, dispatch);	
– Field service;	
– Maintenance; etc.	**Important Notes:**
	To conclude our discussion, let me add that people have tendency to fall back in the old track.
C.B. (Under **IMPLEMENTATION**)	Implementation needs close follow-up till the change becomes a routine practice. Then only can one be sure that the new ideas have been fully implemented.
3. FINAL REPORT.	
Distribute Handout No. CPS-06/XII.	Final Report should reflect the value of your efforts and contribute to the common fund of knowledge for the benefits of others in your organization, now and in future.
REPORT ON PARTICIPANT'S PROJECT Based on the Guidelines in the above Handout, ask one of the Group Leaders to prepare a draft report on one of their problems and present it to the whole group in the last session,	
	Before closing this session, may I request one of the Group Leaders to prepare Final Report for "presentation" in the last session.

13

TRAINING TECHNIQUES OF CREATIVE PROBLEM SOLVING

SESSION – XII

– TEST YOUR CREATIVITY

– SEMINAR

Procedure	Statement
TEST YOUR CREATIVITY By Parice Horn	
Distribute Appendix No. AP-12/*XII*. (First 2 pages only) Allow max. 30 minutes. **Distribute Appendix No. AP-12/*XII*.** (Page 3- Answers) **Distribute: CPS-06 & CPS-07/*XII*** **Distribute Annexures- 15 & -16/*XII***	On completion of the test, please count your score and judge for yourself 'Creative Quotient.'

SEMINAR/PRESENTATION OF REPORT

SEMINAR:

If the CPS programme is organized by a Company, a seminar should be arranged as the valedictory event.

(Trainer to Announce Details of Seminar for Benefit of Participants)

1. The name and designation of the Chief Guest/CEO of the Company (and also of other dignitaries).

   – Time.

   – Venue.

   – Sequence of presenting 'reports' in the seminar.

   – Prize distribution (if any).

2. Invitation cards giving the details of programme should be distributed/e-mailed at least one day in advance. Personal invitations to Chief Guest and two "Judges" is recommended.

3. Ensure that the two eminent persons (judges) who maybe the Heads of Departments, are available for evaluation of the "Presented Reports."

4. Also, meet the Chief Guest' and apprise him of the workshop deliberations. (Supply key points to him. If possible, provide a draft for his address.)

5. Spend 10 minutes with each Team to understand what & how they would like to present Reports? What aids? Who is going to present?

Procedure	Statement
6. Explain the sequence of presentation to the Team Leaders.	

7. Announce procedure:
 	
    Who is presenting?
 	
    What he should do?
 	
    Time limit.	

START PROGAMME:
 	
– Welcome address by the Trainer.
 	
– Presentation of Reports by the Team Leaders
 	
– Address by Chief Guest.
 	
  (Distribute copies of address immediately after presentation is over.)
 	
– Announce prizes.
 	
– Prize distribution.
 	
– Vote of Thanks
 	
  (Script to be prepared earlier).
 	
CLOSE SESSION
 	
HIGH TEA	

IN LIEU OF SEMINAR, RESENTATION OF PARTICIPANT'S PROJECT REPORT Based on the Guidelines given in Handout No. CPS—12/XII, ask one of the Group Leaders to prepare a draft/final report on one of their problems. Ask the Group Leader to "present" the report to you as the Panel Leader. You may play the role of Management Representative and put some interesting questions to ensure lively debate. Close the CPS programme by thanking the Participants for their enthusiastic interest in the course, and with the hope that the inputs will be of great help in improving their Creative out put.	

14

APPENDICES
HANDOUTS
ANNEXURES

APPENDICES

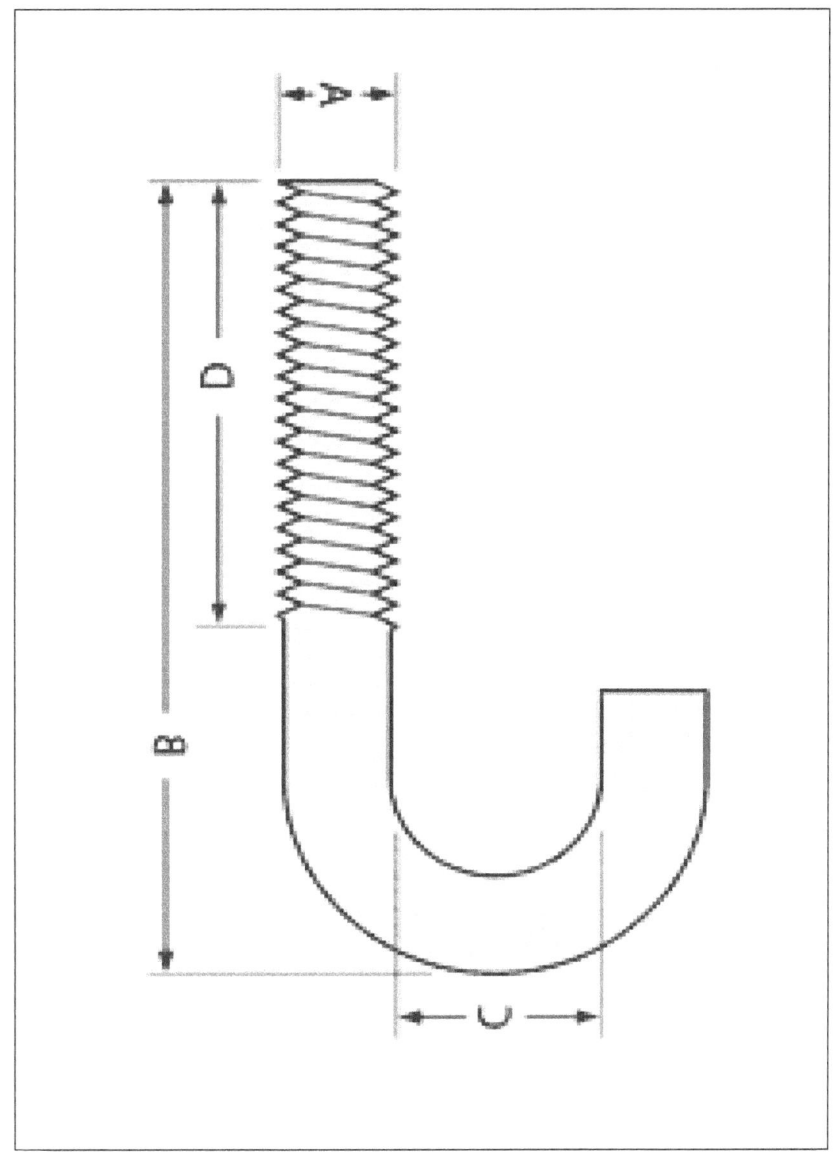

NB: This is a sample picture. There are hundreds of pictures available on Internet. Search 'TREES' or 'BIRDs' on Google and select one of your Liking.

SIXTY SEVEN WAYS TO STOP CREATIVITY

- A good idea but - - - -
- Against company policy.
- Ahead of the times.
- All right in theory - - - -
- Be practical.
- Don't be ridiculous.
- Can you put it in practice?
- Costs too much
- we can't pay for the tools.
- Don't start anything yet.
- Have you considered - - - -
- I know it won't work.
- It can't work.
- Doesn't fit human nature.
- It has been done before.
- It needs more study
- It's not budgeted.
- It's not good enough.
- It's not part of your job.
- Let me add to that - - - -
- Let's discuss it.
- Let's form a committee.
- Let's make a survey first.
- Let's not step on toes.
- Let's put it off for a while.
- Let's sit on it for a while.
- Let's think about it some more.
- Not ready for it yet.
- Of course, it won't work in our Department.
- Our plan is different.
- Our business is different.
- Some other time.
- That's not our problem
- The boss won't go for it.

- The new men won't understand.
- The old timers won't use it.
- (Can't teach old dog new tricks)
- The timing is off.
- The Union won't go for it.
- There are better ways.
- It will create Industrial relation problems
- Too academic.
- Too hard to administer.
- Too hard to implement.
- Too late.
- Too soon.
- Too many projects now.
- Too much paper work.
- Too old-fashioned.
- We have been doing it this way for a long time and it works.
- We haven't the man power.
- We haven't the time.
- We're too big.
- We're too small.
- We haven't done it (that way) before.
- We've tried it earlier - - - -
- What bubblehead thought that up/
- What will the customer think?
- What you're really saying is - - - -
- Who do you think you are?
- Don't you have better things to do?
- Who else has tried it?
- Why hasn't someone suggested it before If it's a good idea?
- You are off base.
- Surely you know better.

....

Adopted from: "The Management Of Intelligence" – Greg

SHOT BLAST CHAMBER

CONVENIENT CLAMPS

PROPOSAL:

In a large Equipment manufacturing company, having about 70 numbers of conventional milling machines, jobs are clamped using U-strap clamps… Open ends of clamps rest on jobs and closed ends are supported on stacks of assorted materials consisting of blocks, pieces of flats and sheets. In selection of appropriate mix by trial needs lot of patience and entails avoidable waste of machine time in setting up the jobs.

Manager of Technology Innovation Workshop manufactured and used "Self-Supporting."

Clamps illustrated in the accompanying sketches. Set up time was reduced by 50% to 60%.

This idea was conveyed to 'J & F' and Tool Design department, with request to procure such clamps and supply to all Milling Machine Centres.

RESPONSE:

- The very idea of providing 'Convenient Clamps' is in line with internationally followed system, – BUT the respondent points out some lacunae:
- Bolt with pad, nut and lock nut have become extra items.
- Can be used only whenever the job height is smaller.

  [Please note: (1) Subjective statement (2) About 60% jobs are 'smaller.']
- Lock Nut is redundant.
- Welding of nut to U-clamp 'spoils' it, and can't be used for other jobs.

  [Why should modified clamps be used for 'other' jobs?]
- Finally he suggests use of 'more versatile clamping systems' and provides drawings of clamping arrangements (appearing in some catalogues.)

NB: After writing to Idea Owner and circulating his response throughout the Plant, nothing was done to procure or to make the 'versatile' clamps.

CONVENIENT CLAMPS

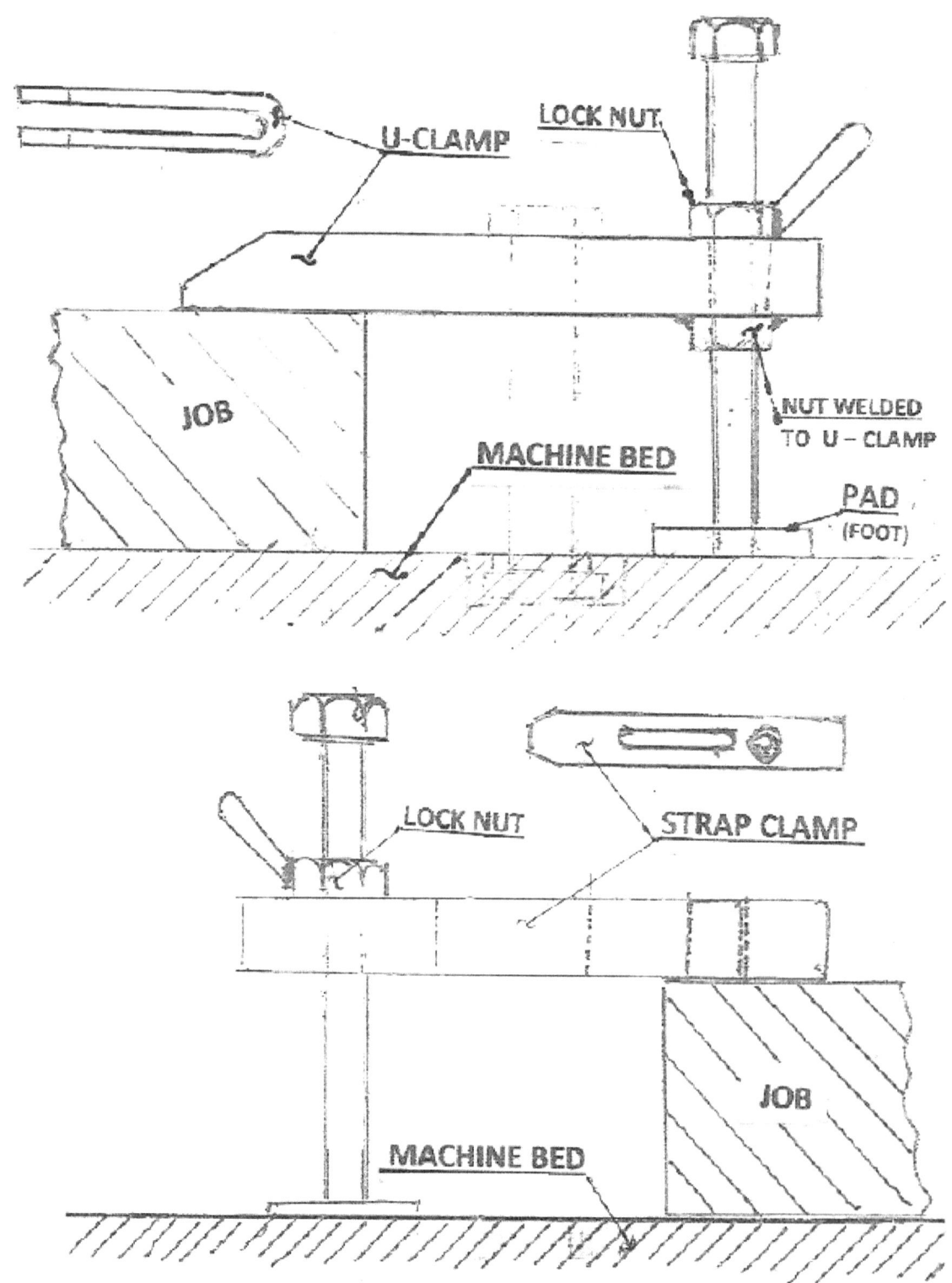

END CLOSURES OF PIPES

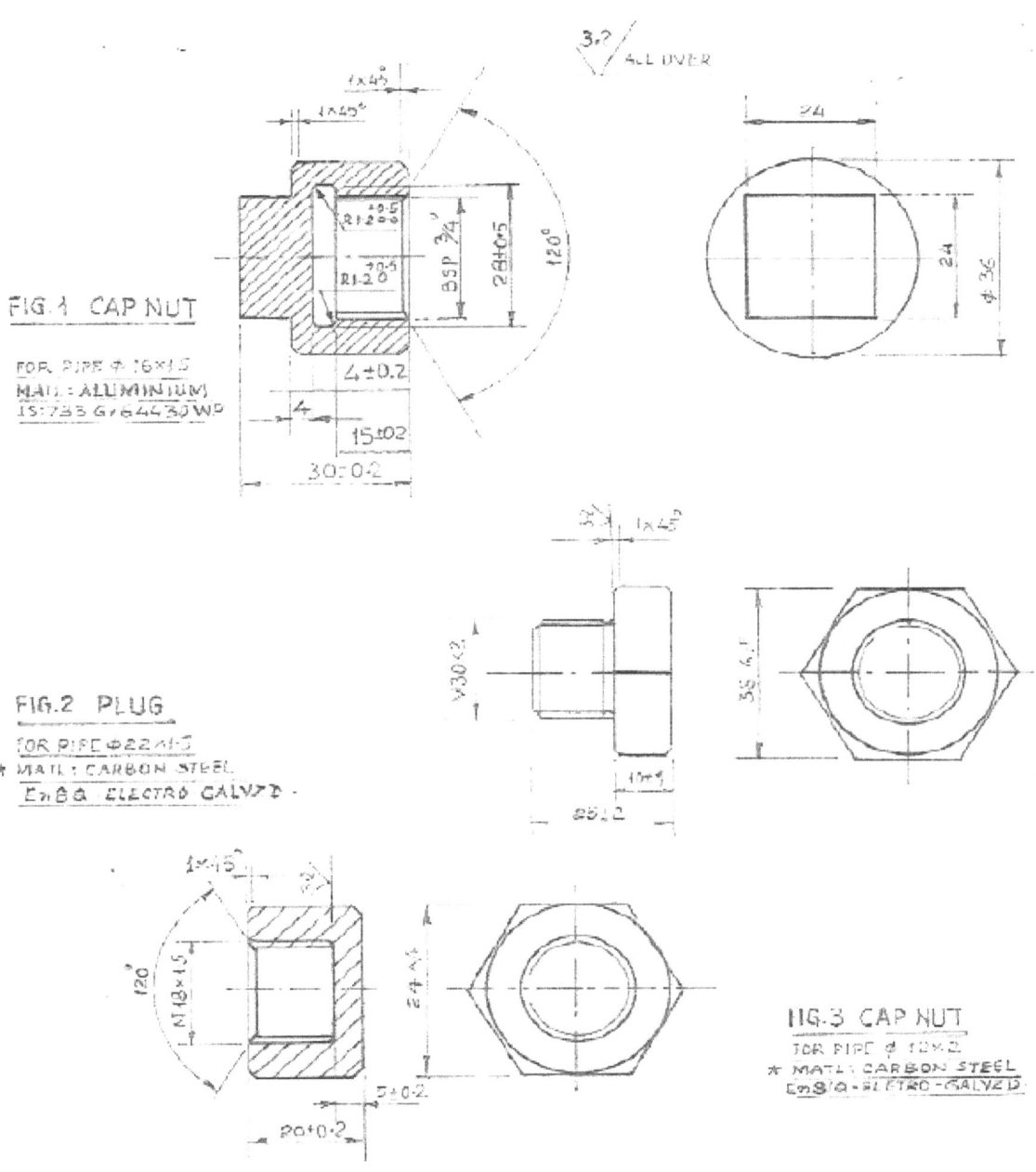

FIG.1 CAP NUT

FOR PIPE Φ 16×1.5
MATL: ALUMINIUM
IS:733 G/64430 WP

FIG.2 PLUG

FOR PIPE Φ22×1.5
* MATL: CARBON STEEL
En8Q ELECTRO GALVZD.

FIG.3 CAP NUT

FOR PIPE Φ 12×2
* MATL: CARBON STEEL
En8Q-ELETRO-GALVZD

NOTE:1. THESE ITEMS ARE REQUIRED DURING DESPATCH AND TRANSPORT
TO CLOSE PIPE ENDS. THE SAME TO BE REMOVED AT THE TIME
OF ERECTION AND COMMISSIONING OF THE 'PRODUCT'

* 2. MATERIAL FOR ITEMS 2 AND 3 SUBSEQUENTLY CHANGED TO ACUMINUM
SQUARE BARS 32 AND 24 SQS, RESPECTIVELY.

3. THERE ARE TOTAL 7 SIZE OF WHICH TYPICAL 3 ARE SHOWN HERE

4. QTY.: 4 OF EACH SIZE (TOTAL=28 PC) PER 'PRODUCT'.

ANSWERS TO PROBLEM IN EXERCISE NO. 53

USES OF SAND: Silica (SiO_2), commonly called quartz. Silica comprises 65% of Earth's crust.

Common Uses:

1. Popcorn (Sand bath)

2. Holder for liquid bottles.

3. Fire extinguisher.

4. Dreamers shape it into castles, pyramids (Sand Art).

5. Children play with it.

6. Our forefathers used it in 'Hour Glasses.'

7. Filtering of water.

8. Ballast for railway tracks.

9. Construction: 95% of sand goes in construction field: Buildings, roads, bridges, pre-cast Structures, dams, etc. etc.

10. Bending of steel pipes without wrinkles, Blasting to remove rust, self defense, sand bags for closing breaches in tanks and dams, as counter weights, etc.

After transformation and modification:

Feldspar, Mica, basalt (that form black beaches of Hawaii), Granules of coral (that form the pink beaches of Bermuda), Man-made fly ash, steel slag – all contain quartz.

1. Glass (3/4th sand): Think of thousands of items made of glass.

2. Metal castings (ranging from water taps to engine clocks and much more).

3. Crystals of quartz that can be made to vibrate at a constant, pre-determinable rate, depending on their size and shape.

4. Quartz crystals in transmitters (Radio and short waves transmission, T.V. etc.)

5. Quartz crystals in modern watches.

6. Chains of silicon and oxygen is a new group of synthetics called "silicones" produced as Plastic, rubber, fluids and gelatins.

7. Silicones find applications in polishing wooden furniture, shield of space-crafts to non-stick cooking wares.

8. In medical field, silicones are used for body parts (finger joints, nose, breast, heart valves, etc.)

9. You will find most sophisticated application of silicon in Electronics:- Silicon chips are the basic building blocks of calculators, computers, Microprocessors and all modern devices.

....

Solutions to problem in Exercise No. 54

1. $1 + 2 + 3 + 4 + 5 + 6 + 7 + (8 \times 9) = 100$
2. $(-1 - 2 + 3 - 4) \times (5 - 6 - 7 - 8 - 9) = 100$
3. $-(1 \times 2) - 3 - 4 - 5 + (6 + 7) + (8 \times 9) = 100$
4. $1 + (2 \times 3) + (4 \times 5) - 6 + 7 + (8 \times 9) = 100$
5. $(1 + 2 - 3 + 4) \times (-5 + 6 + 7 + 8 + 9) = 100$

And some not so obvious ways:

6. $1 - 2 - 3 + (4 \times 5) + 67 + 8 + 9 = 100$
7. $1 + (2 \times 3) + 4 + 5 + 67 + 8 + 9 = 100$
8. $(-1 + 2) \times (34 + 56 - 7 + 8 + 9) = 100$
9. $1 \times 2 + 34 + 56 + 7 - 8 + 9 = 100$
10. $12 + 3 - 4 + 5 + 67 + 8 + 9 = 100$
11. $123 - 4 - 5 - 6 - 7 + 8 - 9v = 100$
12. $123 + 4 - 5 + 67 - 89 = 100$

....

DOCUMENTS FLOW CHART

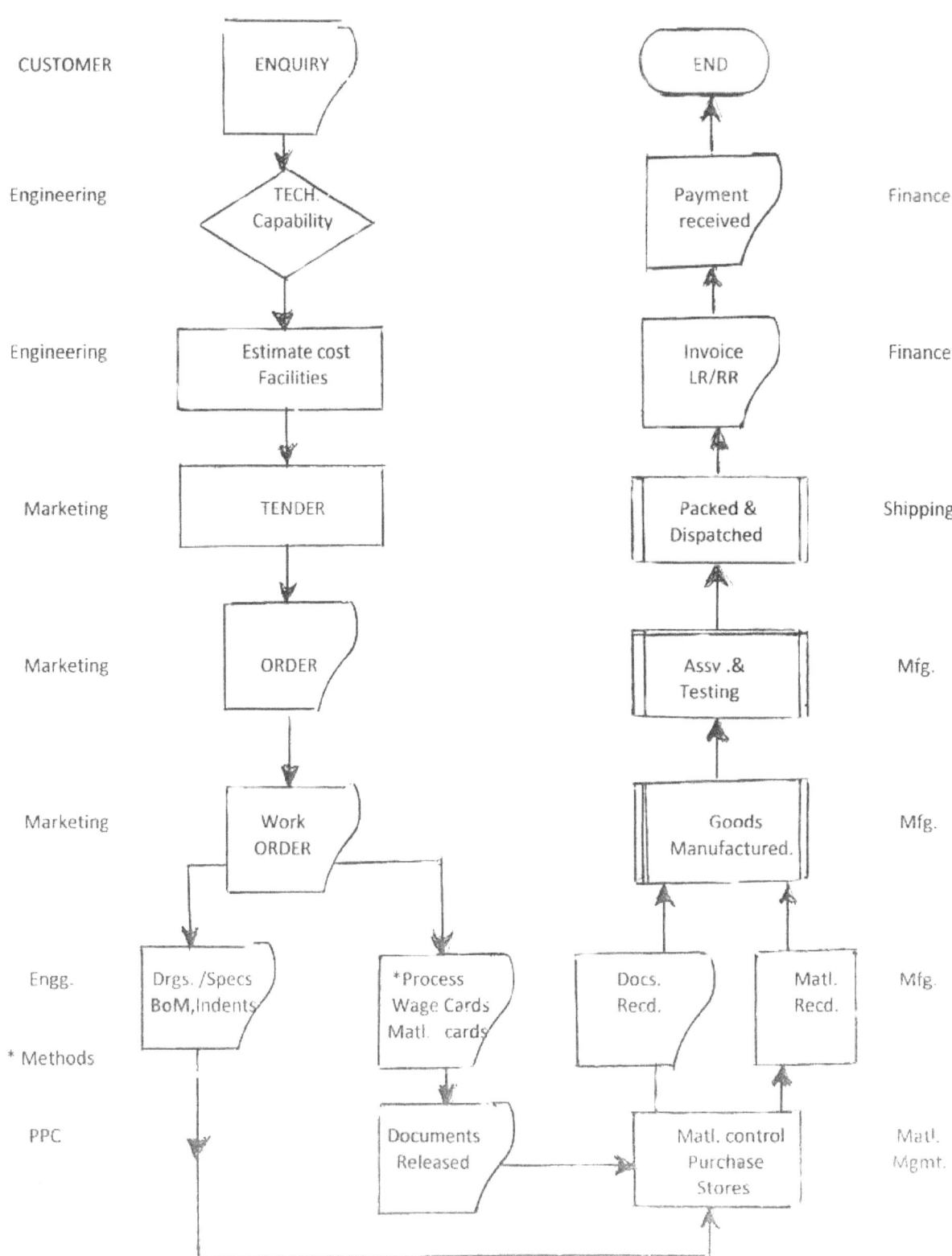

INFORMATION/DATA COLLECTION

* **MARKETING/COMMERCIAL:**
    - Customer's requirements
    - (Reliability, Serviceability, etc.)
    - Market demands.
    - Marketing strategies.

* **DESIGN/ENGINEERING:**
    - Dimensioned drawings (of parts and assemblies)
    - Latest specifications (fits, finish, tolerances, etc.)
    - Weight of parts.
    - Operating conditions. (Max. & Min. Temps. pressures and other environmental factors.)

* **PRODUCTION PLANNING & CONTROL:**
    - Scheduling.
    - Capacity (make or buy decision).
    - Batch size.

* **METHODS ENGINEERING:**
    - Process specifications (Machine tool, J & F, Heat and Surface treatment, etc)
    - Material preparation plans. (Material utilization, cutting allowances, etc.)
    - Work standards/Operation times.
    - Level of skill.

* **MATERIALS MANAGEMENT:**
    - Suppliers (Current & Potentials)
    - Purchasing cost
    - Vendors performance & development.
    - Delivery schedules.
    - Inventory carrying costs.
    - Max, and min, stocks and anticipated arrivals
    - Storage, handling, distribution, disposal, etc
    - Scrap records.
    - Cost of bought-out items.

* **PRODUCTION/MANUFACTURING:**
    – Process constraints.
    – Most likely delivery date.
    – Deviation and re-work.
    – Available skills.
    – Models or prototype of parts or complete set of dismantled components.

* **QUALITY CONTROL & TESTING:**
    – Incidence of rejection & rework.
    – Third party quality requirements.
    – Inspection and testing procedures.
    – Limits of deviation allowed.

* **COSTING/FINANCE:**
    – Product cost in terms of materials, labour, overheads, etc.
    – Cost of rework & rejections.
    – Full analysis of work costs.
    – Value added/Conversion cost.
    – Return of Investment.

* **PACKING AND DESPATCH:**
    – Packaging
    – Handling
    – Transportation
      = Ware-housing
    – Distribution/Delivery of finished goods.

* **MISCELLANEOUS:**
    – Maintenance requirements
    – Safety requirements
    – In case of consumables;
    – Performance data
    – Consumption pattern
    – Environmental influence
    – Special care in storage, handling, disposal, shelf-life, etc.

Ex.

1. Cutting oil causes dermatitis; hence it should be non-corrosive to job and machine tool components, and non-toxic/non-irritant.

2. Some materials like cutting oils & lubricants should remain stable during service life, and be compatible with each other.

3. Dust of H.G.L. & F.G.R.P. and fumes of acids, paints and other chemicals are injurious to health.

DESIGN OF A PEN HOLDER

Normally, a house-wife keeps a ball pen or pencil in a kitchen shelf cluttered with other articles like cans, bottles, utensils, etc., and beyond the reach of small children, Consequently and invariably, the pen gets 'lost' or misplaced and it takes efforts to trace it on the shelf.

A group of Technician apprentices fresh from polytechnic were asked to help the lady by suggesting some means to keep the pen in pre-assigned place. The only condition imposed was that the pen should not be tied by thread and hung from a nail, since it is required at many places in the kitchen.

Problem	:	By consensus, the problem was worded as "Design a Pen Holder" suitable for a single ball pen (with Clip).
Function	:	"Hold pen" (primary)
		Ensure stability
		Provide safety (secondary)
Suggested Alternatives	:	Out of 41 ideas generated in 15 minutes, 50% were of conventional pen stands of various designs & materials usually found on the tops of office tables, and the remaining 50% had some element of novelty.
		These suggested 21 alternatives are illustrated in the enclosure (Page 155)
Your tasks	:	1. Prepare a list of factors or criteria for evaluating relative merits of the alternatives.
		2. First Team will subject the first 7 ideas to Criteria T- Chart technique and give their decision to 'accept' or 'reject' or 'hold on.'
		3. Second Team will take the next set of 7 ideas and arrive at a decision after quantifying the criteria expressed in absolute or percentage form. (Criteria Value-T Chart).
		4. Third Team will tale up the balance ideas and on the above lines, using one of the 2 evaluation techniques.

Finally, the whole group will decide the 'winner,' using Value-T-Chart.

PEN HOLDER

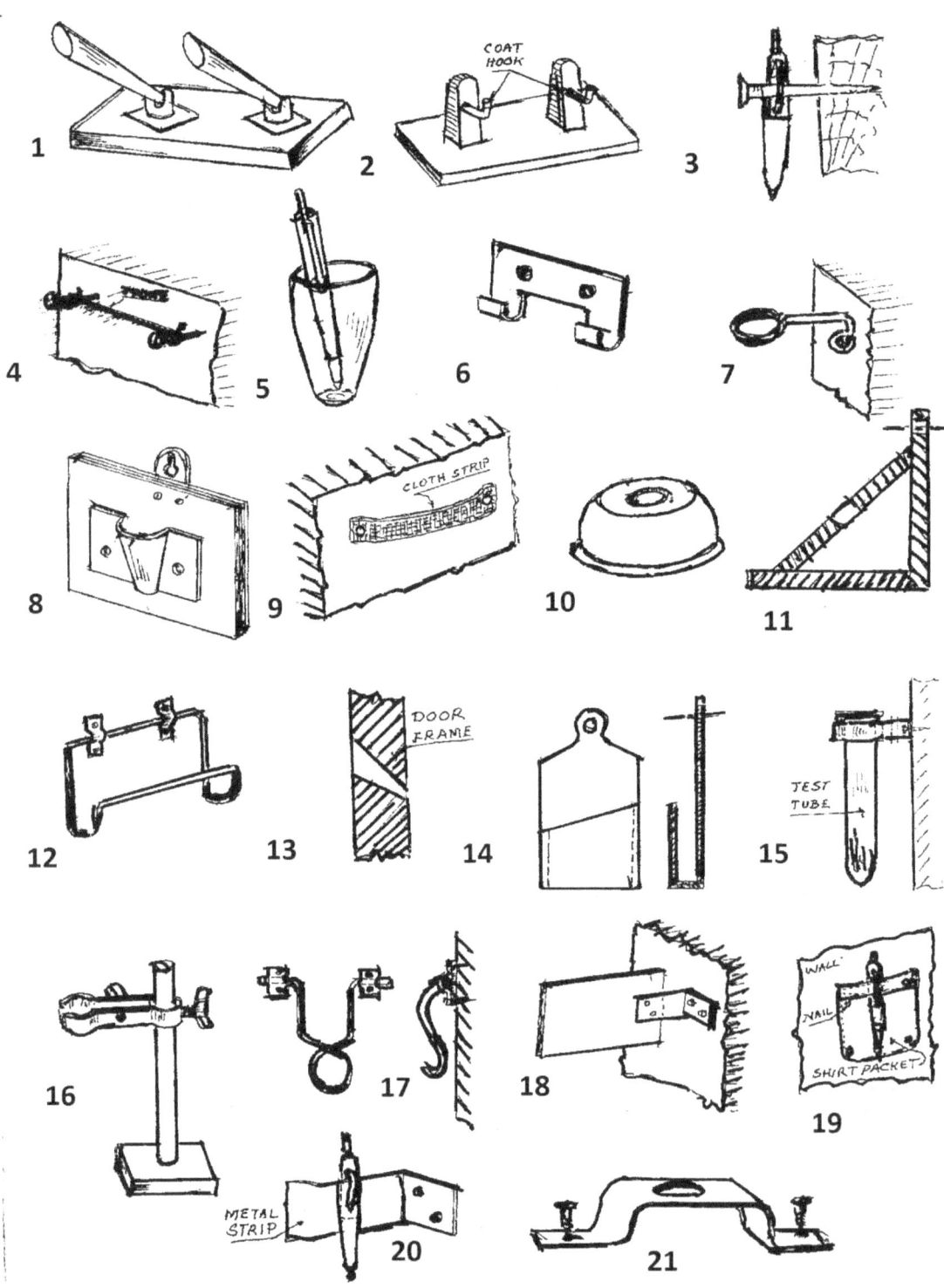

TEST YOUR CREATIVITY

By Patice Horn

Psychologists use many tests to measure creativity—which is often defined as combining the original, forming new combinations of old elements.

Dr. Sarnoff Mednick of University of Southern California, devised a remote associations test similar to this one. He found that highly creative people can complete this exercise more easily than non-creative people.

Please note:

This test is based on Western adages/quotes/proverbs. Therefore, don't take to heart if you do not score above five.

Continued/2

TEST YOUR CREATIVITY

How creative are you? Fill in each blank with a word that the three given words have in common. For example, answer to the first question is "bath" creating "bath robe," steam bath, bird bath.

				Answer
1.	robe	steam	bird	(bath)
2.	book	post	tour	( )
3.	fly	shoe	radish	( )
4.	dew	bee	comb	( )
5.	bag	coffee	stalk	( )
6.	key	wall	cold	( )
7.	leaf	berry	window	( )
8.	up	hot	life	( )
9.	down	stone	golden	( )
10.	cake	blue	cottage	( )
11.	hut	skirt	roots	( )
12.	table	court	elbow	( )
13.	line	light	dress	( )
14.	end	shelf	maker	( )
15.	hole	brake	servant	( )
16.	port	pool	cable	( )
17.	line	busy	spelling	( )
18.	fore	bow	painting	( )
19.	day	house	weight	( )
20.	made	cuff	left	( )

Score ( )

ANSWERS TO QUESTIONS IN CREATIVE TEST

1. bath
2. guide
3. horse
4. honey
5. bean
6. stone
7. bay
8. line
9. touch
10. cheese
11. grass
12. tennis
13. head
14. book
15. man
16. car
17. bee
18. finger
19. light
20. hand

Handouts

(Forming Parts of Course Material)

IT COULDN'T BE DONE

(Edgar A Guest)

Somebody said that it couldn't be done.

But he with a chuckle replied

That 'may be it couldn't but he would be one

Who wouldn't say to till he tried.

So he buckled right in with the trace of a grain

On his face. If he worried he hid it

He started to sing as he tackled the thing

That couldn't be done, and he did it.

Somebody scoffed "you'll never do that"

At least no one has ever done it.

But he took off his coat and he took off his hat,

And the first thing we knew he'd begun it.

With a lift of his chin and a bit of a grin

Without any doubting or quid it;

He started to sing as he tackled the thing

That couldn't be done, and he did it.

There are thousands to tell you it cannot be done.

There are thousands to prophesy failure,

There are thousands to point out to you one by one,

The dangers that wait to assail you.

But just buckle in with a bit of a grin,

Just take off your coat and go to it:

Just start in to sing as you tackle the thing

That "cannot be done," and you will do it.

PROBLEM SHEETS

I. "Sense" or "Smell" Types:

1. Worker asked for grievance procedure and Forms.

2. **An employee in anger banged the door after him.**

3. Just before the arrival of dignitaries, public address system started behaving erratically.

4. **Your son is appearing for the Public Examination.**

5. Water-works employees announce strike of works.

II. "Anticipate" or "Size-In-Advance" Types:

6. Railway track is submerged in flood water. Trins are likely to be cancelled.

7. **Shift Timings announced to be changed, merging general shift with first shift.**

8. Order Book position gives alarming picture for the next year.

9. **Only 10% of employees are to be promoted on merit basis and 60%, on carrier growth basis.**

10. Manufacture of new product requires a spherical turning tool.

III. "Imposed" or "Come-To-you" Types:

11. **Works complain that tea is not served on work spots.**

12. On charge-sheeting a worker, other workers resorted to 'Go Slow' tactics

13. **Union delegate approaches you with 'request' to grant 'French' leave to the whole section to attend cradle ceremony of their colleague's first child.**

14. A lady (neighbor) from Non-vegetarian family requests an orthodox vegetarian family to lend pressure cooker.

15. **<u>An employee from shop floor request you to post him general shift to enable him to attend evening classes.</u>**

IV. "Invited" or "Run-Into" Types:

16. A crane operator loyal to Asst. Foreman, shouted at and refused to obey Foreman's order. Department watching.

17. **An Office Superintendent (O.S.) exhibited air of superiority on getting promotion to Officers cadre. Later, he requested his O.S. to look after his duties so that he may go on leave. O.S. refused, retorted back, Section watching.**

18. In order to ensure perfect 'health' of your TV set during the ensuing Cricket season, you opened it and cleaned it inside out but now it has gone dead.

19. **Production Engineer asked the machinist to deburr machined jobs. The worker refused saying that it was Fitter's job.**

20. You took your P.A. to task before other staff for not being in his chair. Later, it transpired that on previous day, you have asked him to go to H.Q. for getting your note approved before coming to the office.

V. "Search The Truth" Types:

Great Scientists, Inventors and Philosophers suffered untold miseries in search of truth.

21. Siddhartha (Gautam Buddha) earlier to renouncing his throne, was a king, had all the worldly pleasures at his disposal – wealth, health, power of a king, beautiful wife and a child. Still he was the most unhappy man in search of truth: Why the man is greedy? Why a man hates man? Why a man kills man? Why the man torments, etc.

22. **Nicolas Copernicus, the great astronomer had known the staggering truth "concerning the revolution of Heavenly bodies" since he was a youth. But the book he wrote on the subject at the age of 40 was withheld from publication for more than 30 years of hesitation in the face of threat of Church. He suffered imprisonment and torture as a heretic.**

23. Galileo, the great experimentalist assembled his first telescope and put it to use despite opposition from the Church. The orthodox astronomers refused to look through it branding it a gadget which creates the vision of devils. Finally, the Church banned the use of telescope and Galileo faced the Inquisition. The poor, blind and helpless Galileo suffered mental and physical torture for 9 long years in the prison and then he died.

24. **A.L. Lavoisier, the father of modern Chemistry, the Inventor of decimal system of coinage and measurement and a great philanthropist, was born in a wealthy family and was th friend of King Louis of French. He had all creature comforts at his disposal but in search of scientific knowledge, he suffered immensely. Hounded to his death by the jealousy of a few fanatics. He was finally beheaded by the same King who was his friend, at the young age of 51.**

25. Faraday's electrical generator was the product of his abiding faith in the spirit of inquiry. He merely wondered what would happen if he were to spin a copper disk between 2 poles of a horse shoe magnate. In pursuit of knowledge and truth, he faced many unpleasant situations, which in common parlance are called probe

----

To Which Of The Above 5 Categories Do The Following Problems belong?

26. Management has decided to introduce hydraulic clamping systems on all important machines.

27. **Scooter runs erratically; stops whenever the owner drives or rides on pillion but runs beautifully when the mechanic or his friend or both of them ride together.**

28. Forging die broken, Components urgently required for assembly and testing. Customer invited to witness performance test.

29. You have been asked to improve productivity of your machine shop by 20%.

    No additional resources will be allocated.

30. **As a Q.C. Engineer your pet theme in every meeting and communication was 'bad finish' of all products. You are now posted in "Finishing Department"**

31. 80% of rejection of shafts before assembly and 50% of the remaining in field service are due to induction hardening of journal portion. You are asked to study and recommend remedial measures.

32. **LPG gas leaking from the valve. Lapping can stop the leakage for 2 or 3 months.**

    Replacement takes one month, says company.

33. Heat treated a job with several tapped holes in it., without reading the process. Threads distorted. Taps broken while re-tapping the threaded holes.

34. **Spectacles forgotten in the office. Old pair of glasses without frame available in the house. Giving decision on report and notes urgent.**

35. Wife asked you to leave office one hour earlier and pick up kids on way back. Important meeting is scheduled from 4.30 P.M.

NB: Trainer should make attempts to add many more examples from his own experience and narrate 4 or 5 in each programme.

****

THE CALF PATH

– Sam Walter Foss.

One day through the primeval wood,
a calf walked home as good calves should;
But made a trail all bent askew, a
crooked trail as all calves do.
Since then three hundred years have field,
and I infer the calf is dead.
But still he left behind his trail,
and thereby hange my moral tale.
The trail was taken up next day,
by a long dog that passed that way;
And then a wise bell-weather sheep,
And draw the flock behind him, too;
as good bell-weathers always do.
And from that day, o'er hill and glade,
as good bell-weather always do.
And from that day, o'er hill and glade,
through those old woods a path was made,
And many men wound in and out,
and dodged and turned and bent about,
And uttered words of righteous wrath,
because it was such a crooked path;
But still they followed – do not laugh
the first migrations of that calf
And through this winding wood-way stalked,
because he webbled when he walked,
This forest path became a lane,
that bent and turned and turned again;
This crooked lane became a road,
Where many a poor horse with his load,
Toiled on beneath the burning sun,
and travelled some three mile in one.
And thus a century and a half,
They trod the footsteps of that calf.

....2/-

The years passed on in swiftness fleet,
The road became a village street,
And this, before men were aware,
a city's crowded throughfare.
And soon the central street was this,
of renowned metropolis;
And men two centuries and half,
trod in the footsteps of that calf.
Each day a hundred thousand men,
follow this zigzag calf again,
And o'er his crooked journey went,
the traffic of a continent,
A hundred thousand men were led,
by one calf near three centuries dead.
They followed still his crooked way,
and lost one hundred years a day.
For thus such reverence is lent,
to a well-established precedent.
A moral lesson this might teach,
were I ordained and called to preach;
For men are prone to go it blind,
along the calf-path of the mind,
And work away from sun to sun,
to do what other men have done.
They follow in the beaten truck,
and in and out, and forth and back,
And still their devious course pursue,
to keep the path that others do.
They keep the path a sacred groove,
along which all their lives they move;
But how the verse old wood-gods laugh,
who first saw the primeval calf.
Ah, many things this tale might teach
but I am not ordained to preach.

....

THE THINGS THEY SAY ABOUT

I. INSPIRATION:

The creative person is both more primitive and more cultivated,
more destructive, a lot madder, and more saner, than an average person.

– Frank Barron

Genius is one percent inspiration and 99% perspiration.

– Thomas Edison.

The imagination imitates. It is the critical spirit that creates.

– Oscar Wilde.

Creativity varies inversely with the number of crooks involved in the broth.

– Bernice Fitzgibbon.

II. ON WISHFUL THINKING:

We would be still in dark age were it not for wishful thinking.
Without it most of the achievements of mankind would have never been started.

– Alex Osborn

– In Make-up Your Mind.

III. ENTHUSIASM:

Philosopher Eric Hoffer enthused:

- I love ideas as much I love women. I derive a sensuous pleasure from playing with ideas. Genuine ideas dance and sing. They sparkle and twinkle with mirth and mischief. They titillate the mind, kindle the imagination and warm the heart. They have grace and pomise.

Gordon Parks, author and film director:

- Enthusiasm is the electricity of life. How do you get it? You act enthusiastic until you make it a habit. Enthusiasm is natural; it is being alive, taking initiative, seeing the importance of what you do, giving it dignity, and making what you do important to yourself and to others.

- **Henry Ford;**
- **Enthusiasm is at the bottom of all progress. With it, there is accomplishment. Without it, there are only alibis.**
- **Enthusiasm: the hardest work.**

IV. <u>MIND:</u>

- Mind is like a parachute. It works only when it is OPEN.
- Mind is like a camera: Objects before camera are recorded only when the shutter is open. Shutters of our mind are called "Panchendriyas" – or 5 senses of organs.

V. <u>ON MIND-TRAINING:</u>

- The only effective way to train the mind is to use it persistently what interest it.
- **Constantly apply mind to problems that interest you and strive to solve them.**
- Back it with resolute will. How do you strengthen will?
- **Like muscles, it grows by exercises. (Like muscles, don't allow it to go flabby and inert by disuse).**
- First set small and attainable tasks. Exercise mind regularly. And by degrees, make the tasks harder and achieve them.
- <u>Reward:</u> **enhancement of self-respect deep inner satisfaction.**
- Maintain the attitude of perpetual student. (Mind is ever young).
- **Training is like a jig – guides the tool which does the work of shaping material – Your tool is practice – How sharp is your tool?**

(R.G. Chaudhari).

VI. <u>ECOMONY OF EFFORTS:</u>

- Don't pick up a cannon to chase a sparrow. Don't use sling shot against a forty-ton battle tank.

VII. <u>EDUCATION</u>

– **Almost all our teaching tends to cramp creativity. A well-filled mind is certainly essential for creativity but stuffing our mind with inert ideas which can never be utilized or tossed into new combination just numbs the imagination – Every stresses the intake and retention of such data.**

– Even in teaching of Arts and Crafts, there is no scope for stimulation of imagination and no room for initiative efforts. Teacher selects the patterns, material, colours, tools, methods and folds, cuts, etc. etc. and all children come out with exactly the same design.

– Same size of brush and same brand of colours… Why not fingers as brush and floor as painting surface?

– 'Plastecin'! Why not potters clay?

– Pre-dotted figures of animal and birds – outlined figures to be coloured

  • Activities should be such as to teach the students to think divergently, laterally – to formulate new mental images and concepts.

VIII. <u>SOME WISE WORDS</u>:

– **Use judicial ability not to curb but to keep imagination on track.**

– Let imagination enlighten judgment.

– **Be hopeful and trust yourself to create ideas**.

– Beware of perfectionism… anything and everything can be improved later.

****

LITTLE RED HEN

No one really knows who wrote this updated version of the well-known fable. But it has been widely reprinted and even read at shareholders' meetings.

Once upon a time, there was a little red hen who scratched about the barnyard until she uncovered some grains of wheat. She called her neighbors and said, "If we plant this wheat, we shall have bread to eat. Who will help me plant it?"

"Not I," said the cow.

"Not I" said the duck.

"Not I," said the pig.

"Not I," said the goose.

"Then I will," said the little red hen, and she did.

The wheat grew tall and ripened into golden grain.

"Who will help me reap my wheat?" asked the little red hen.

"Not I," said the duck.

"Out of my classification," said the pig.

"I'd lose my seniority," said the cow.

"I'd lose my unemployment compensation," said the goose.

"Then I will," said the little red hen, and she did.

At last it came time to bake the bread. "Who will help me bake the bread?" asked the little red hen.

"That would be overtime for me," said the cow.

"I'd lose my welfare benefits," said the duck.

"I'm a dropout and never learned how." said the pig.

"If I'm to be the only helper, that's discrimination," said the goose.

"Then I will," said the little red hen. She baked five leaves and held them up for her neighbors to see.

They all wanted some – in fact, demanded a share. But the little red hen said, "No, I can eat the five loaves myself."

"Excess profits!" yelled the cow.

Capitalist leech!" cried the duck.

I demand equal rights!" shouted the goose

The pig just grunted. Then they hurriedly painted "unfair" picket signs and marched around, shouting obscenities.

The government agent came and said to the little red hen,

"You must not be greedy."

"But I earned the bread," said the little red hen.

Exactly," said the agent. "That is the wonderful free-enterprise system. Anyone in the barnyard can earn as much as he wants. But, under government regulations, the productive workers must divide their product with the idle."

And they lived happily ever after. But the little red hen's neighbours wondered why she never baked bread again.

GUIDELINES FOR PREPARING CPS REPORT

1 INTRODUCTION

In Introduction, a brief mention may be made of 3 categories of problems, namely:

i. Judgmental problems which evoke a mere 'Yes or No' type of answers,

ii. Analytical Problems which yield only one correct solution and

iii. Creative Problems have a large number of solutions, none of them may be absolutely correct but most of them are workable, now or later.

Next, a brief description of the following should be given under this heading:

– The scope of the study, limitations and restrictions.

– Main objective sought to be achieved (in terms of present and targeted costs, Quality, Serviceability, etc.)

– Decisions/Views of the experts.

There are 3 steps in solving such problems: Viz. (a) Problem Finding, (b) Fact Finding and (c) Solution Finding.

2. PROBLEM SELECTION:

A brief mention may be made as to how and why the problem was selected for Creative attack. Some tips are given below:

– Problems suggested by our Team Members were put through the sieves and creative type of problems was taken up for exploration.

3. FACT FINDING (INFORMATION/DATA COLLECTION):

All essential facts collected from authentic sources (such as Applications and Marketing, Engineering, Technology, Procurement agencies. Field services, etc.), should be included under this heading.

If the problem is related to a product or process, reference should be made to specifications, drawings and/or sketches.

4. IDEA FINDING (CREATIVE SOLUTIONS):

There are many techniques of finding ideas. We have followed the simplest but most effective technique for generation of profitable ideas, called Group Brainstorming.

All ideas that are put forth by the members of the group are listed below, with possible exception of those that are known to be seriously in error from the known causes.

(No result or idea should be excluded merely because it is unexpected or is inconsistent with the others or with theoretical considerations. Sometimes, an unexpected result points to further significant experimentation.)

(Ideas/suggestions may be listed concisely in serial order).

5. EVALUATION (SELECTION PROCESS OF IDEAS):

Mention should be made as to how the time interval between idea finding and evaluation session was utilized for collecting more information/data for developing the potential ideas suggested in Brainstorming session(s).

The following points should also be included in this section;

- – How objective assessment of ideas' worth was made during this session.
- – How Judgment of ideas unspoiled by habits, attitudes or pre conceived notions helped in selection of the best from a number of competing solutions.

Brief description of Criteria like performance, economy, quality, number of part (fewer or more), production, marketing, handling, storage, etc. followed in Evaluation, may be included in this section.

6. PLANNING AND EXECUTION (DEVELOPMENT):

After giving a brief account of criteria selected for consideration, a table for each developed idea should be prepared, indicating how the developed idea compares with the existing design on the basis of:

- – Functions performed,
- – Estimated costs,
- – Effects on criteria selected, etc.

Each table must include an estimate of implementation cost and estimated time for implementation.

In some cases, tabular statements may be supplemented by, or the whole section may be, written in narrative form.

7. IMPLEMENTATION

After 'final selection' of developed ideas but before implementation (and preparing the final report), presentation may be arranged to gain acceptance, giving

- – Present and proposed design details,
- – Implementation cost,
- – Action plan, etc.

----

8. REMARKS:

Remarks against developed ideas are mostly in the following forms:

- Discussed with specialists, suppliers, or maintenance, etc.

- Dropped/Rejected as… consumption is likely to go up or deviates from the Present policy or it is not in conformity with customer's requirement, etc.

- Potential idea requires fuller development.

- Try out to be arranged.

- For future development.

9. ACKNOWLEDGEMENT:

Acknowledgement of help, co-operation and encouragement received from persons in CPS may be given on first page of the report.

ENVIRONMENTAL FACTORS

1. Stimulating environment is one in which there is always something to do new, to experience and to know – and where one is called upon to response to new tasks and challenges.

   The monotonous environment or task dampens creativity.

2. Encouraging and rewarding environment stimulates innovation and experimentation.

3. Environment that ignores creative efforts or penalizes them, dampens it.

   Environment that does not evoke defensiveness or fear of criticism encourages creativity.

4. Environment of constructive criticism and where there are opportunities
   For feedback foster creative efforts.

5. Environment that provides opportunities for technical training, gaining of
   First- hand experience and scope for experimentation, stimulate creative efforts.

6. Environment that provide freedom of thoughts and action coupled with responsibility and accountability encourage creativity – and authoritarian environment stifles creativity.

7. Love, admiration and status bestowed on the innovators and creators stimulate enthusiasm and encourage creative efforts.

****

ANNEXURES

(FOR THE USE OF TRAINER ONLY.)

NOT FOR DISTRIBUTION AMONGST PARTICIPANTS OF CPS COURSE)

ADDITIONAL EXERCISES TO BRING OUT PROGRAMME PHILOSOPHY

(In place of exercise No.3/Session-I)

1. How do you put a pumpkin in a pitcher?

   (Position the pitcher in such a way that a small growing pumpkin enters its mouth.

   Allow the pumpkin grow inside the pitcher fer 3 months)

2. While removing label from the empty glass bottle, it slipped out of my hands.

   What might have happened to the bottle?

   (All ideas are likely to hover around the pull of gravity -- -- bottle breaking or not breaking.

   But no one is likely to say that if it were released in a space-craft, it would remain where it was released or if it were released under water, it would go up if it mouth was closed with a cock or go down if its mouth was open)

3. *A 'VESPA' type scooter was found to stop after running for about 200–250 metres whenever the wife of owner used to ride on pillion, but not when the owner and mechanic were riding together, for any distance. What could be the cause(s) of this funny phenomenon?

   (Polythene tube between the fuel tank and carburetor was replaced. Instead of cutting the tube to exact length, the whole length of 2 metres was coiled and left in the cavity of there fuel tank which was getting folded and stopping the flow of petrol whenever a heavy person was sitting on the pillion.)

4. *A scooter was found to stop after running for 1 to 1.5 Km. The carburetor and ignition systems were found to be in perfect condition and there was enough petrol in the tank. What could be the possible cause(s) of this phenomenon?

   (The Engineer owner had replaced the rubber gasket in the fuel tank cap by a rubber disc, thereby blocking the air vent.)

ADDITIONAL EXAMPLES TO ILLUSTRATE THAT MORE PROBLEMS ARE CREATED IN SOLVING A SINGLE PROBLEM

1. Describe the sequence of events following someone stepping on a 'NAIL', which was invented to facilitate joinery work.

   (Once upon a time, wooden planks were so cheap and nails were so precious that people used to burn down their houses to get back the nails intact.)

2. Another variation can be generated by narrating the sequence of events following a tyre burst caused by a nail.

3. *Consider an example of Factory Safety Act and Leave Rules for Factory Workers.

   The Safety Act defines a reportable accident as one in which the worker is away from work for more than 48 hours. Industries are anxious to reduce reporting of such accidents. On the other hand, as per Leave Rules, leave is granted if the worker is unable to perform his duty.

   Now, take the example of a worker who is unwilling to work but doesn't want to apply for leave. He approaches the supervisor with a tightly wound bandage on his thumb soaked in petrol, pretending that he had a deep cut while working on his machine, and orally request for 3 days leave.

   The supervisor avoids application of Safety Act and allows the worker to sit idle. The worker triumphantly comes out of cabin, goes to his work centre and boast of ingenuity.

   And to celebrate his 'victory', offers cigarettes And strikes a match stick, and lo and behold!!. The petrol soaked bandage on his thumb caught fire. The rest is very clear.

   Very interesting trains of events can be woven around the topics of:

4. Internet.

5. Institution of marriage

6. Contraceptive pills

7. Noise pollution.

8. Antibiotics

And thousands of inventions, Laws, Rules, etc. Only sky is the limit in describing the episodes- to bring out that solution to one problems invariably leads to more problems.

ADDITIONAL EXAMPLES TO BRING OUT DEFINITION OF PROBLEM

1. Management has decised to introduce hydraulic clamping systems on important machines.

2. LPG gas leaks from the valve. Lapping can stop leakage for about 2 months. Replacement I takes one month, says the company.

3. Forging die broken. Components urgently required for assembly and testing. Customer invited for performance test.

4. 80% rejection of shafts before assembly and 50% of the remaining in field service are due to induction hardening of journal portion. You are asked to study and recommend remedial measures.

5. Computers unauthorized used by watchmen after working hours. Damage reported. Officer says: Do some thing to prevent the misuse.

6. You have been asked to improve productivity of your machine shop by 20%. No additional resources will be allotted.

7. As a QC Engineer your pet theme in every meeting and communication was 'bad finish.' Of all products. You are now posted in 'Finishing Department.'

8. Heat-treated a job with several tapped holes in it, without reading process. Threads distorted. Taps broken while re-taaping the distorted holes.

9. Spectacles forgotten in the office. Old pair of glasses without frame are available in the house. Giving decision on report and notes urgent.

10. You are racing to catch up with time for an important appointment, but the vehicle tyre is punctured.

11. In the middle of movie on TV, fuse is blown off.

12. In the presence of your Factory Inspector, cap of gas cutting equipment blown off.

13. Half way on way to movie your wife gets nagging doubt whether she had out off the gas stove or not.

14. On reaching railway station you start wondering whether your pet cat is inside the house or whether the water taps are open.

Use 2 or 3 examples in each programme.

ROUTINE PROBLEMS AND ROUTINE SOLUTIONS

1. What is the time now?	–	Look at the watch/Or ask some one having a watch.
2. Tyre is flat.	–	Change it or inflate it'
3. Child is refusing to go to school	–	Coax, cajole, bribe or beat him.
4. Water tap (faucet) is leaking?	–	Replace washer.
5. Boss is angry?	–	Pamper his ego.
6. Ink pen leaking?	–	Apply soap on threads.
7. Sighted a scorpion?	–	Kill it.
8. Nut is jammed?	–	Soak it in kerosene.
9. Indisposed?	–	Go to doctor.
10. Cook is absent?	–	Eatm in hotel/Resort to self help.
11. Expecting (unwanted) guests?	–	Go somewhere as guest.
12. Want a spouse?	–	Get married.
13. No money?	–	Beg, borrow or earn.
14. Cloths are dirty?	–	Wash them.
15. Wife is angry?	–	Praise her 'beauty',
Can you add 5 more examples?	–	If yes, add to this list.

………………………

JUDGMENTAL/LOGICAL PROBLEMS AND DECISION TYPE SOLUTIONS

Solutions to such problems rely on your judicial ability in evaluating and choosing between: yes or no; true or false; moral or immoral; logical or illogical; etc.

1. Shall I buy a new car?

2. Should I go to court or seek compromise?

3. Shall we have an early dinner?

4. Should the Chairman be elected or nominated?

5. Shall I buy gold or invest in stock (shares)?

6. Should we use ceramic tiles or polished stone slabs?

7. Weather is coudy; shall I carry an umbrella?

8. Should the capital punishment be abolished?

9. Shall be make or buy parts?

10. Should the old parents be put in "Home for the Aged"?

11. Why not abolish all coins?

12. Is it rational to levy income tax on salaries?

13. Should we stockpile nuclear weapons?

14. Are the inter-cast marriages desirable?

15. Should we import food items like pulses, onions, etc.?

Think of 5 more questions of judicial type and add to this list.

ANALYTICAL PROBEMS AND THEIR SOLUTIONS

Analytical problems can be solved step-by-step (heuristically) and each problem yields only one correct solution. Such exercises strengthen the muscles of mind.

1. Sixteen match sticks are laid down to form 10 triangles as shown in Fig.1
   Remove 4 match sticks and leave 4 separate triangles.

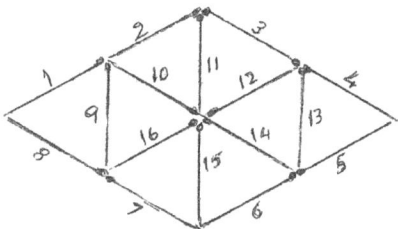

Fig. 1

2. Add 3 straight lines to 3 triangles of Fig 2 to make 13 triangles.

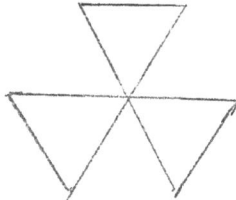

Fig. 2

3. Remove 3 match sticks from Fig. 3 and rearrange to get 3 squares.

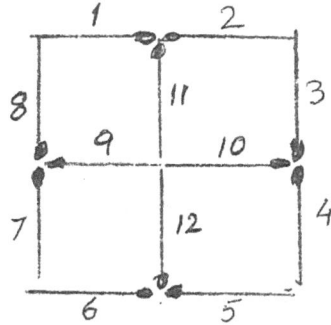

Fig. 3

ANNEXURE – 06/II

4. Make just one cut and arrange to form 'T' as shown in Fig. 4 on RHS.

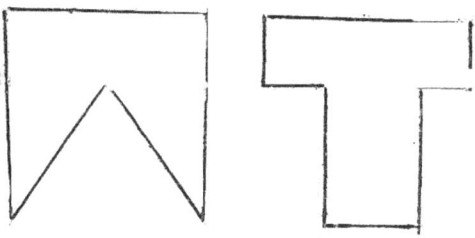

Fig. 4

5. There are 25 pots containing 5 varieties of flowers, 5 pots of each variety.

   Arrange the pots on a chequered floor having 25 squares as shown in Fig. 5, in such a way that no flowers of same variety appear in a line, vertically, horizontally or diagonally.

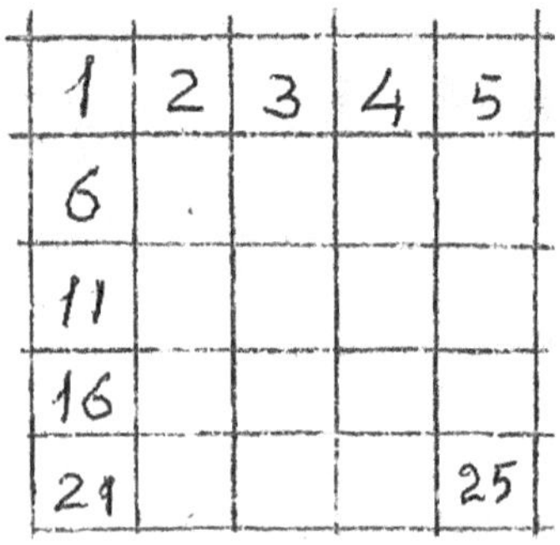

1
2
3
4
5
6
11
16
21
25

Fig. 5

6. Dial of a wall clock in Fig. 6 is cracked into 3 pieces in such a way that the sums of digits in 3 sectors add up to 12, 12 and 36 Show how?

12
1
2
3
4
5
6
7
8
9
10
11

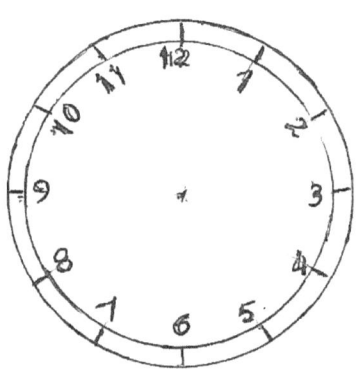

Fig. 6

7. By adding just 2 more straight lines to the following Figure, form 3 arrows.

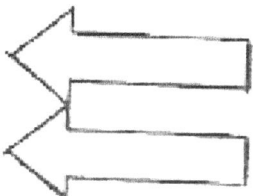

8. Develop a general formula for numbers of which the last digit is 5.

9. You are given a rimless spherical pot with a small opening at the top and half filled with rice. You are also given a square hand kerchief diagonal of which is smaller than the largest circumference of the pot.

   Describe a most comfortable method of carrying the pot with rice without making any changes in the given materials and without touching the pot with fingers while carrying it.

10. Add 5 more match sticks to the following 6 to make 9.

11. Find the odd-man out:

    Hammer, Chisel, Screw Driver, Bolt, Handsaw, Spanner.

12. While walking through a thick jungle, you have come to a 3-path (Y) junction, where the road bifurcated into 2 paths. One of them leads to a village of a friendly tribe and the other one to the village of hostile, cannibal tribe. People of both tribes are identical in all respects. They look alike, dress alike and speak the same dialect. The only difference is that the people of friendly tribe always speak truth and the cannibals always tell lies. As you were worried and wondering as to which path you should take, a tribal approaches you offering help but on one condition. You should ask one and only one question. You do not know the tribe to which he belongs.

    What could be your question?

ANSWERS TO PROBLEMS IN ANNEXURE-06/I

1. Remove (a) 3, 12, 7 and 16. (b) 2, 10, 14 and 6.

2. Draw lines AB, AC and BC.

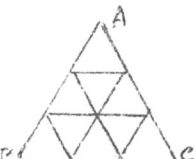

3. Remove dotted lines 2, 3 and 6 and arrange like this.

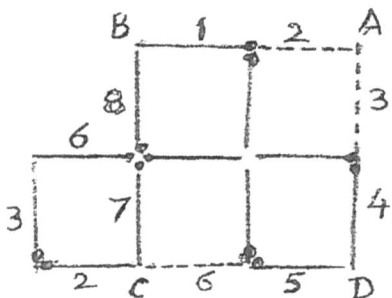

4. Cut along line AB and rearrange 1 and 2 as shown on RHS.

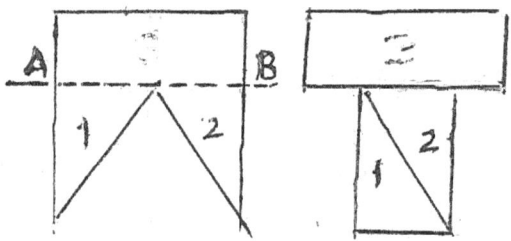

5. Many alternatives are possible. One is shown below.

   {Actually, it is a creative problem}.

A	C	B	E	D
C	E	D	B	A
D	B	C	A	E
B	A	E	D	C
E	D	A	C	B

6.

7. **Draw** lines AC and BC

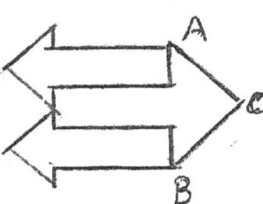

8. {5a}^{2} where 'a' is an odd number.

9. Take out rice from the bowl. Spread kerchief on the bowl covering its mouth. Press down the kerchief a little and pour rice grains in the trough, holding 4 corners of kerchief in hand. Form a bundle inside the bowl and carry it.

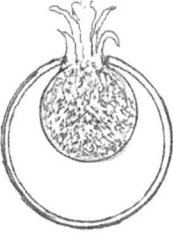

10. Additional 5 match sticks are shown by dotted lines.

11. **Bolt.**

12. **Which road leads to your village?**

CREATIVE PROBLEMS AND CREATIVE SOLUTIONS

Creative problems have a number of solutions, none of which is absolutely correct but most of them are workable now or later.

1. Think of all possible uses of empty tooth paste tube.

   {Replace Paste Tube by Hawaii Slippers or Old socks, Gitar, Egg, etc.{Any one per batch.}

   I made erasers, tap washers, as silencers for dining chairs and many more uses of old slippers after washing with soda and kerosene.

   I used eggs, custard apple, etc, for making moulds in Plaster of Paris to make more eggs, etc. This hobby helped me in making models of patterns of complicated Valve Bodies and other industrial products.

2. Think of all possible methods of finding percentage of water in milk.

   {For example: Deep freezing, evaporation, measuring density, bio-filter, etc.}

3. Suggest many ways of holding the parting lines of a shirt together.

   {e.g. buttons, single stitches, hooks, buckles, zippers, safety pins, clips, pins, verlo straps, etc.}

   Trainer should look out for some unusual/novel method.

4. In how many ways can you raise level of water stored in a plastic jar?

   {Solutions: Pour more water or any other liquid, put sand or stones, etc. but rarely will someone may suggest "put some foating material or press the bottom of jar up}

5. You are given two identical metallic pieces, A and B and told that one of them is a magnet. You are in a glass case containing no other objects.

   In how many ways can you identify which one is a magnet?

   {Solutions:

   a. Keep 'A' on the floor and run one end of 'B' along the length of 'A.' If B is a magnet, A will be attracted by B at all points.

   b. In the reverse test, if B is a magnet, A will not be attracted by the midpoint of

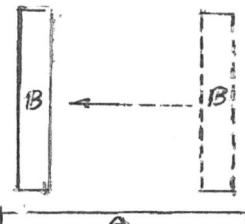
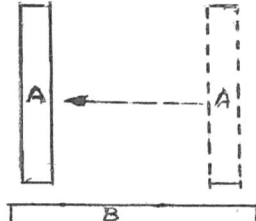

c. Arrange the two pieces as shown in the following Figs, and move the horizontal piece towards the vertical piece till a perceivable attraction is felt.

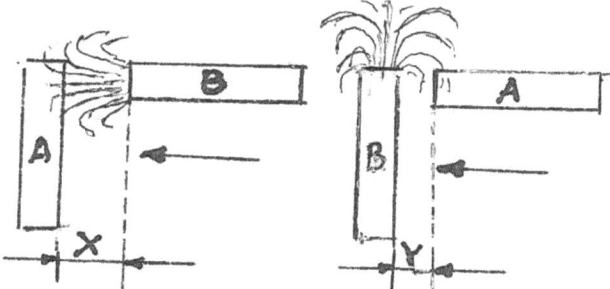

If B is a magnet, X is much greater than Y for the felt attraction.

d. Break one of the objects into 2 pieces and test them with each other and also with unbroken object.

e. Go on dropping A for a month{!} and test it with B. If there was no attraction between them, A was a magnet, and got demagnetized.

f. Try magneto-therapy with both pieces.

   If still you are not satisfied with number of solutions,

g. Pluck you long hair {if not bald} – no pun intended – and suspend A and/or B freely. Whichever gets oriented in North-South direction is a magnet.

h. Keep one the pieces on a wooden block floating on water in a tray. Magnet gets oriented in N_S direction.}

6. You are given 2 Mild Steel rings of different diameters.

   In how many ways can you join them together rigidly, using any material and any process.

   {Solutions: Usually the solutions will be in the form of two concentric rings connected together by 2 or 3 or 4 rods welded radially to the rings as shown in Set-1 of Figs.

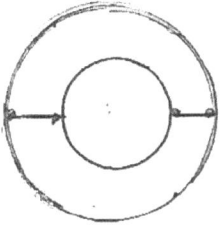 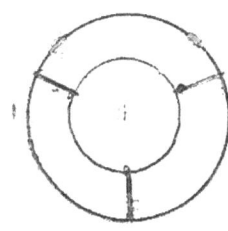 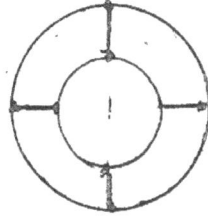

SET-1

Comments: The limited number of solutions shown in SET - ! Are the result of

Conditional thinking and too many assumptions.

Continuing Exercise No. 6, very rarely someone would suggest:

i.   Smaller ring outside or inside the larger ring, as shown in Fig.-a and Fig.-b.

ii. Rings connected by 2 straight rods as shown in Fig. c or 2 rods as shown ib Fig.-d.

SET-2

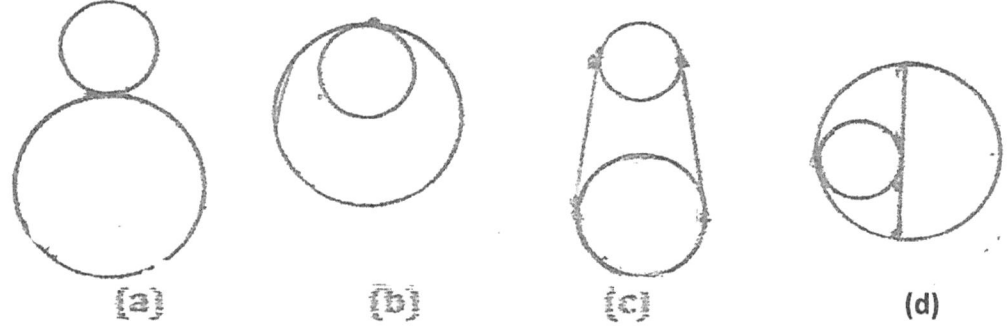

(a)   (b)   (c)   (d)

iii. Rings connected by curved or zig-zag lines, even by 1 or more straight lines or circles.

SET-3

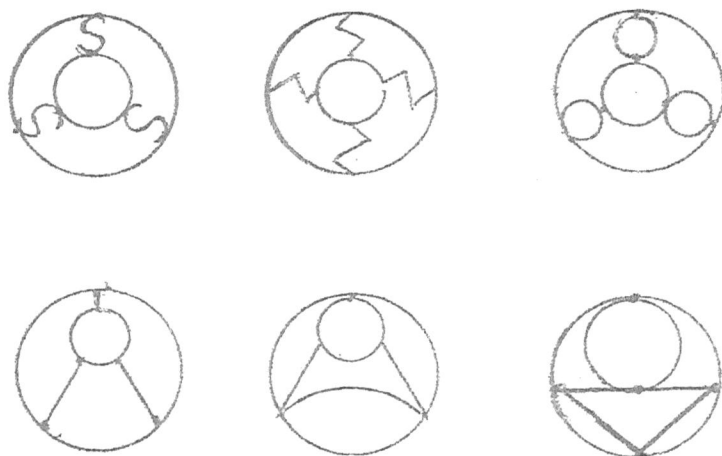

Comments: Had you only freed your mind from restrictions or "concentricity" and Straight, radial rods, you would have got infinite solutions.

ADDITIONAL EXERCISES TO BRING OUT "ABSORPTIVE" AND "RETENTIVE" ABILITIES

1. Exhibit pictures of 2 or 3 familiar objects like:

   Car, Cow, Hammer, Pitcher, Clock, etc. and ask the participants to identify them.

2. Play back some sound recorded on your mobile phone:

   Sound of temple bells, popular song from a movie, Tabla, Cuckoo, etc. {any one} and ask the participants to identify the source of that sound.

3. Pass on a small can/bottle containing white crystals of:

   a. Rock Salt or

   b. Alum

   Announce that it is an edible substance and that their job is to name it

4. Carry 5 to 6 common objects like Marbles, Ampoules, Tea Spoons, Chalk Pieces,

   Erasers, etc. in a cloth bag. Invite 2 or 3 participants {one by one} to come to the head of your table and identify the objects by putting hand in the bag.

   Note: The objects should be smooth, without sharp corners or edges.

5. Draw a brief word picture of:

   a. Chilli Powder.

   b. Pomeranian Puppy.

   c. "Veena" or any other musical Instrument.

ADDITIONAL EXERCISES ON "REASONING" ABILITY

{It is advisable to record the proceeding on video and play it back while explaining how the debate went on.}

1. Suggested topics for discussion are given below. The participants have to select any one topic and discuss its various aspects for about 15 minutes.

    a. Managements policy on handling Union Demants.

    b. Language policy of Central Government.

    c. Holiday pattern in Government offices.

    d. State of affairs in sports.

    e. Working of Police department.

    f. Reservation policy on cast basis.

    g. Abolishing of currency notes.

    h. Any other topic making headlines today.

2. Prepare a questionnaire of the following type and let the participants record their reactions/ responses against each item in 2 or 3 words.

    Your Response

    a. After seeing your wife in a new dress.   ---------------------

    b. After 'interviewing a girl with matrimonial intension.'   ---------------------

    c. After listening to an election speech   ---------------------

    d. After visiting an exhibition of paintinfs.   ---------------------

    e. After enjoying an Indian Marriage Feast.   ---------------------

    f. After witnessing a ghasty accident.   ---------------------

    g. After entering a witness box in court   ---------------------

    h. After testing a newly prepared mango pickle.   ---------------------

    i. After driving on a badly maintained road.   ---------------------

    j. After meeting your Income Tax Officer.   ---------------------

3. If you are given a chance, will you marry a rich widow of your age? If yes, why? If not, why not? Answer in 4 or 5 words.

4. If you become a dictator of a country, what would be your priorities? Mention any 5 areas.

5. Mention four major ground of frequent quarrels between you and your spouse.

6. Rank the following attributes in order of their importance from your point of view.

		RANK
a.	Obedient	{ }
b.	Accepts judgment of peers.	{ }
c.	Courteous	{ }
d.	Sophisticated	{ }
e.	Does work on time.	{ }
f.	Self assertive.	{ }
g.	Always asks questions and questions.	{ }
h.	Visionary.	{ }
I.	Determined.	{ }
j.	Independent in judgment.	{ }

{NOTE: First five traits are farthest from, and next five are closest to creative personality.}

....

ADDITIONAL EXERCISES TO DEMONSTRATE CREATIVE ABILITY

1. List the objects that are red in colour.

   {Note: Chidren in age group of 6–8 first listed Mummy's Sari(dress), teacher's lipstick, bangles, bindi (Tilak), apple and rose, items to which they are constantly exposed and belonging to persons whom they adore.

   Children in age group of 14–16 had different priorities: Blood. Post box,

   Lord Hanuman, Railway porter, cricket ball, flag, etc.}

2. Think of objects that are:

   a. Pink, soft and edible.

   b. Smooth, cold and repulsive.

3. Think of all possible excuses for rejecting a marriage proposal from received from rich parents of a most beautiful girl.

4. What could be the possible reasons for a jet-set executive to mary a fat, dark and ugly woman?

5. Lists as many words as you can that end in TV.

   (The catch is VT = vity as in creativity.)

6. What are the interesting and unusual uses of:

   a. Old bicycle tube, (I used it as a door closure)

   b. Coconut shells. (I used them as patterns for hemi spherical castings)

   c. Charcoal (I sprinkled fine powder on water in a tray and drew a floating Rangoli' on the powder.)

   d. Pieces of copper wire. (Draw line sketches of object, cut and bend copper pieces to match lengths of lines, and solder joints. Hang your 'ART' on wall)

7. What are the possible uses of thumb?

    {In addition to daily uses, lead the group to think of its uses in gesticulations: "Mudras," teasing, show of strength, 'failed' sign and also in tracing nail pictures, thumb impression, crime detection, etc. (Can you write, eat food, hold things without thumb?)

8. The only vehicle (a car) that I possess for commuting in the 'town' has gone for servicing. How do I go to office?

    (After getting usual responses, disclose the solution.)

    Solution:

    a. I live on the 12th floor and my office is on the ground floor of the same building -- Perils of presumption, strait-jacket thinking of travelling in horizontal plane only).

    b. Today is Sunday or I am on leave.

9. In how many ways a medal or identity card be worn (held) on the chest pocket of your shirt?

10. You and your wife are on "no taking" term for the past one week.

    Think of as many ways as you can to 'break the ice' without losing your face.

    (The most interesting suggestion that I came across is as follows:

    Husband puts off all lights and start searching 'some imaginary thing' everywhere, making some queer sounds with objects – like dragging steel trunk, running comb over the edge of an empty can, etc.

    Wife: What are you searching at this odd hour?

    Husband: Your voice, darling.

FOOD FOR THOUGHT

1. Rita was a mod, charming young lady of liberal views, always exuding mirth and warmth. However, on returning from her honeymoon trip, her friend Lina found that the merriment has vanished and day-by-day Rita was getting morose and gloomy. Finally after some weeks, Lina enquired about reason for such a sudden and unexpected turn of events. Rita took a deep breath and confined in her:

   "It is Robrt's- eversince I have come back from honeymoon, he has not even kissed me once--"

   Sympathizing with her friend's plight, Lina asked advisingly: "Then, why don't you seek divorce from him?"

   Problem: What would have been the possible response of Rita?

   (Answer: How can I, when I am not even married to him?)

2. Mr. Masta Ram, flam-buoyant son of a tycoon was admitted to a nursing home. After knocking the door, a beautiful young lady doctor entered his special room, closed the door behind her and ordered him to undress and lay on the coach for check up. After examining him thoroughly, she explained him the nature of his ailment and the line of treatment he was to undetgo. Before leaving, she enquired: "Any Question?"

   Mr. Masta Ram asked "-------------?"

   Problem: What would have been Mr. Masta Ram's question?

   (Answer: Well, why did you knock the door, in the first place?)

EXERCISES TO BRING OUT ABILITY TO FANTASIZE

1. Think of as many adventures a man can indulge into?

   (Marriage is one…)

   Suggest consequences of: (highly improvable events)

2. All human beings ave suddenly lost their power of imagination.

3. A newly invented medicine prolongs vim and vigour of a youth by 50 years.

4. If the Earth and the moon are connected by means of a hose pipe, what would happen?

5. If all the Chemical Factories are banned?

EXERCISES TO DEMONSTRATE THAT OUR CREATIVE POTENTIAL IS DORMANT
(WE ARE NOT AWARE OF OUR OWN CAPABILITIES)

(Questions are designed to elicit novel responses and unusual solutions.)

1. Ask the participants to think of 'nouns' staring with 'K' or 'L' without writing on paper) and to raise their hands if they can recollect more than 10.

   (A very few hands are likely to go up showing that people underestimate their own potentials.)

   Now, ask them to write down and check the score after about 10 minutes.

   Most of them would score more than 10.

   Similar exercises can be conducted by replacing 'nouns' by 'verbs', 'words,' 'synonyms,' etc. for examples:

2. **Verbs** staring with 'V' or 'U.'

3. **Words** starting with 'A' or 'Z.'

4. Synonyms for words 'OLD' or 'GOOD.'

5. Names, in any language, of

   a.   birds– 10 Nos.

   b.   Flowers – 15 Nos.

   c.   Hand tools – 25 Nos.

6. Arrange 5 coins in 2 lines such that each line has 3 coins.

   Now, move one coin and place it in such a way that there are 4 coins in one line and still 3 coins in the other line.

   (Solution: 5 coins are first arranged in either 'L' or 'X' or 'T' or 'V' shape.

   It will be a sort of miracle if someone succeeds in finding the correct solution. Demonstrate how easily the task could be achieved…

   One of the end coin is placed on the top of the coin which is common to both lines.

   Point out that if, and only if they were thinking outside the 2 dimensions in which the 5 coins lay and inside 3 dimensions, could they solve the problem… Don't impose restrictions that won't exist.

Keeping the above solution in mind, you can devise several alternatives, with or without imposing any condition regarding the initial configuration (i.e. 'L', 'V', etc. shape) formed by 2 lines.

7. Arrange 'n' (odd No., say, 7) coins in 2 lines so that each line has {(n + 1) ÷ 2} (i.e., 4 coins) in each line. Then move 1 coin so that there are {(n + 1) ÷ 2) + 1}.

   (i.e. 5 coins) in one line and still (n + 1) ÷2 (i.e. 4) coins in the 2nd line.

   Identical exercises with 'EVEN' No. of coins can be developed as shown in the following examples.

8. Arrange 8 coins in 2 lines – 5 in one line and 4 in another. Next, move one coin and place it in such a way that there are 5 coins in each line.

9. In the above example replace 8, 5, 4 and 5 respectively by 10, 6, 5 and 6

10. By adding 1, or 2 lines, straight or curved, make as many recognizable sketches as you can, from:

S

Some of the possible solutions are shown below:

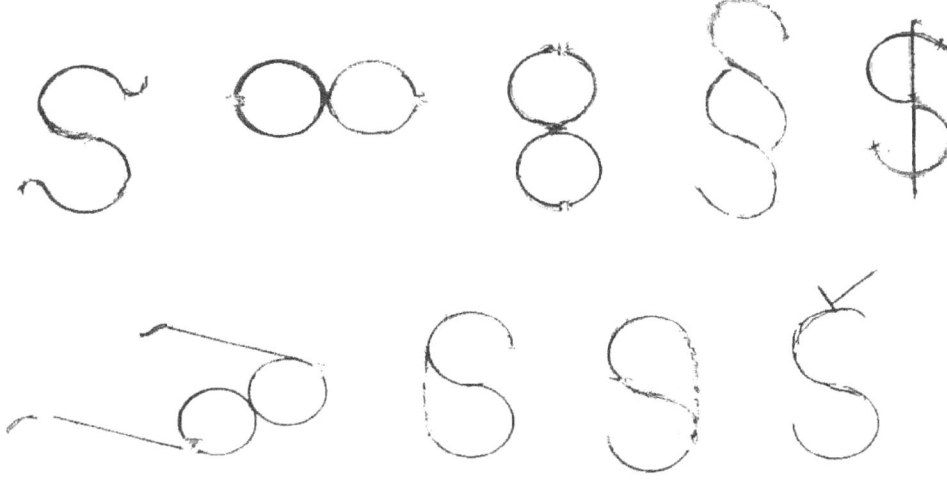

EXERCISES ON 'WORK SIMPLIFICATION'

1. Write "Seventeenth December" on chalk board and ask the participants to write 'this date' as many times as they can in 2 minutes

   Solution: Invariably, people will repeat word by word, but solution could be–

   a. 17th December,

   b. Dec. 17

   c. 17–12- year in 4 digits,

   d. 17/12/year in 2 digits, etc

2. In a knock-out Table Tennis Tournament, 11 players are participating.

   How many Games (Matches) must be played to complete the tournament?

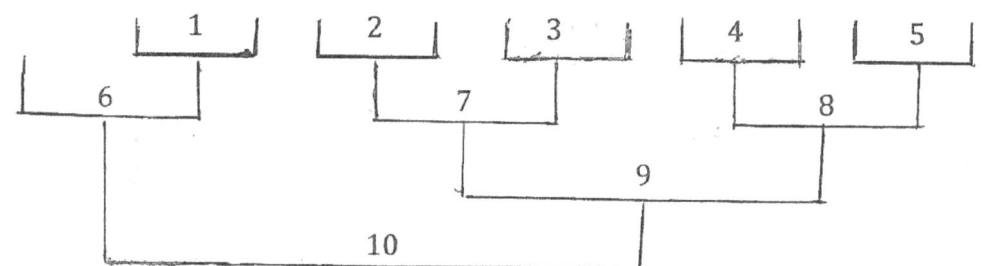

   Number of games to be played = 10

Comments: You have taken (   ) minutes to work out the solution.

If there were 213 players, you would have probably taken (   ) hours to work out the solution.

A novel method of looking at the problem is to think of the number of 'losers' (and of one winner). Since there can only be one winner, there must be 10 losers and to decide 10 losers, 10 matches must be played.

If there are 213 players, 213 – 1 = 212 games need be played.

EXAMPLES OF RIDDLES AND PUZZLES THAT HELP IN EXERCISING MENTAL MUSCLES

1. How many cuts are to be made to divide a metre long paper ribbon into 8 parts?

    Common reply will be 7 cuts.

    (Answer: <u>Two cuts</u>. Fold the ribbon thrice over and cut off the folded ends.

    Each piece will be less than $1000 \div 8 = 125$ mm long. The problem did not say whether the parts should be of equal length of 125 mm.

    Message: Do not impose unnecessary constraints.)

2. There is one word which cannot be spelt correctly. Which is that word?

    (Answer: *Incorrectly*)

3. Form a square (or a cube) using 4 match sticks of unequal lengths. (Cross section of matches is square.)

    (Solutions: There are 2 solutions:

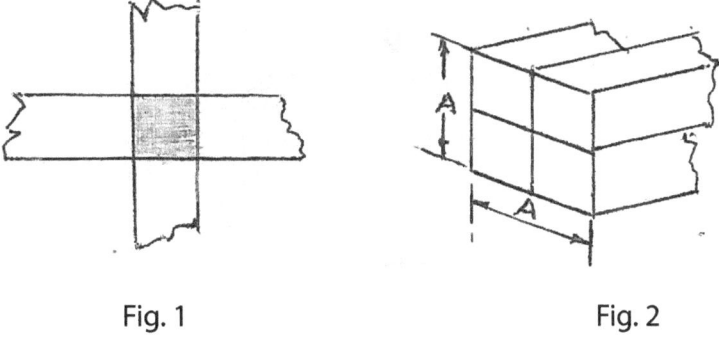

Fig. 1	Fig. 2

    In second solution (Fig.2), the centre is also a hollow cube}

4. Taking "CUBE' in numerical sense, form a cube using 5 match sticks

Solutions:

$$I^{\geqslant} \{1^3\}, V III \{2^3\}, Z7 \{3^3\}$$

SOLUTIONS TO EXERCISE No 35 – (a)

{Divide a square into 4 parts of equal area of any shape.}

(A) A few conventional solutions are shown below.

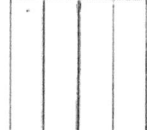 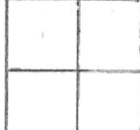 

Straight lines – identical shapes

(B) Non-conventional solutions are shown in the following 3 rows.

(*i*) Curved lines – Dissimilar shapes

  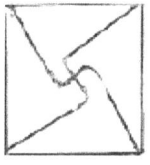

(*ii*) Straight lines – Dissimilar shapes

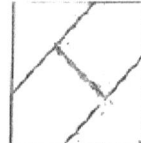  

(*iii*) Identical shapes

BIBLIOGRAPHY

1. Techniques of Value Analysis   Mc Graw Hill/USA
    and Value Engineering -
        by L.D. Miles

2. Techniques of Producing Ideas.   British Publication.
        - James W. Young.

3. Art of Thought – Wallas   "

4. How to Think up   Alex. F. Osborn.

5. Your Creative Power   "

6. Wake up your Mind   "

7. Applied Imagination   "

8. Product Innovation – Knut Holt   Newness – Butter Worths
                                                  Mgt. Library.

9. Fourth Eye-Excellence through Creativity   A.H. Wheeler & Co. Pvt. Ltd
        - by Pradip N. Khandwalla

10. The Nature of Creativity   Research Journal 2006,
        - Robert J. Sternberg   Vol. 18, No. 1, 87–98

11. The Nature Of Creative Development   Stanford University Press
        - Jonathan Feinstein

12. Creative Problem Solving   Available on 'Amazon.'
        by Robrt A.Harris

13. (14) CPS Techniques   Western Kentucky University
        by Jonnes Higgins

14. The OK Boss by James M., (15) How To Be Your Best Friend by Newman & Berkowitz and (16)
    Success Through Transactional Analysis, are worth reading.

For further reading, go to Google and ask for books on Creativity, Creativity in Wild, Creative Problem Solving, etc.

FILMS ON CREATIVITY

1. Why Man Creates?
2. Nature Of Creativity.
3. A Series of Exploration, Episodes and Comments On Creativity.
4. Fooling Around.
5. The Process.
6. The Judgment.
7. A Parable.
8. A digression.
9. The Search.

FEEDBACK

Programme : Creative Problem Solving.

Dates :

We reckon that this feedback from you is very useful for us to improve the quality of our future programmes.

Please feel free to be as frank as possible in your responses.

You need not sign this form, if you so wish.

1. How useful did you find this programme? (Please put a tick mark in the circle.)

POOR	FAIR	GOOD	VERY GOOD	EXCELLENT
2 - O	4 - O	6 - O	8 - O	10 - O

2. Name 2 aspects of this programme which you found very useful:

   (a)

   (b)

3. What are your 2 most important suggestions to improve this programme further?

   (a)

   (b)

4. Will you recommend this programme to your Management/Company?

   (a) Yes, and now - O   (b) Yes, but later - O   (c) No. - O

5. If you want us to facilitate the same, please give us the name(s) of decision making

   Authority(ies) along with his (their) name(s), designation(s) and e-mail/postal address(s). You may attach a sheet.

6. Would you like make any other comments on this programme? Please write below.

   You may attach a sheet.

Name & Signature (Optional)